Private Tutor

SAT Math 2013-2014
Prep Course with Amy Lucas

For additional test prep coaching, contact Amy at www.testpreptutor4you.com.

While every effort is made to ensure this manual is complete and error free, mistakes may occasionally occur. Any errors discovered will be posted in an errata sheet available at www.PrivateTutorSAT.com. Please let us know if you find an error by emailing info@PrivateTutorSAT.com. No teaching tool can guarantee specific score results and Private Tutor makes no such claims.

Written by Amy Lucas
Layout and design by Kyle Broom

All inquiries should be addressed to:
Private Tutor
15124 Ventura Blvd., Suite 206
Sherman Oaks, CA 91403
tel: 818.508.1296 • fax: 818.508.9076
info@PrivateTutorSAT.com

More SAT tutorial books and dvds are available at www.PrivateTutorSAT.com.

PRINTED IN THE UNITED STATES OF AMERICA
9 8 7 6 5 4 3 2 1

Table of Contents

PRIVATE TUTOR SAT Prep Course on DVDs & Books
Praise from Reviewers, Librarians, Teachers, Tutors & Parents

…this series in critical reading is an excellent way to prepare for the exam. Host Amy Lucas is young, lively, and engaging **and brings an extensive background in test prep tutoring to her discussions. She helps students through every aspect of SAT critical-reading work through** *an approach that is clear, well organized, and certainly student-friendly.* **Students can work** *at their own pace to learn how to analyze passages, read the questions, build vocabulary, and understand the formats of the test.* **Lucas is always aware of the variety of student learning styles, presenting** *three different techniques* **for dealing with reading passages.** *VERDICT: A very useful SAT package that can aid students on the test and beyond in all reading work. Strong supplementary materials include a student workbook. Recommended for all collections serving junior high and high school students.*

Library Journal, Critical Reading section review, April 2013

In stark contrast to the "teacher in front of a chalk board"…the presentation is personal, conversational, relaxed, and effective.

Peggy Dominy, Liaison Librarian, Drexel University, Philadelphia, PA

Library collections serving high school students as well as homeschoolers seeking instruction, practice, and basic standardized test-taking tips for the Math SAT will find this series an essential purchase.

School Library Journal, Starred review, Math section

This set of two DVDs and a workbook walks the viewer through the writing component of the SAT and offers advice on how to approach the test. Narrator and tutor Amy Lucas is attractive and vivacious… her delivery makes the videos easy to watch.

Rosemary Arneson, University of Mary Washington Library, Fredericksburg, VA

Amy Lucas brings more than a decade of tutoring experience at the highest levels of the tutoring/ test prep industry to this production. Starting with math vocabulary and then offering a discussion of the range of SAT problems from algebra to geometry to functions, this series offers the student the broadest coverage of the SAT math section. VERDICT: Highly recommended for students facing the SATs.

Library Journal, Math section review

The three different techniques for reading based on ability is an amazing idea. I have never seen it addressed this way and it is so important. **The sample readings were excellent… the one on video games was particularly good…** *the answer explanations are great and are key to a student's success.* **If they can understand their error it will help immensely.**

Kevin Murchie, teacher, Garfield Senior High School, Los Angeles, CA

Teens will easily relate to her upbeat and friendly approach…Lucas's mantra to restate information in your own words before answering a question allows students to trust their judgment, making it less likely that they will be confused by tricky questions.

School Library Journal, Starred review, Critical Reading section

Lucas divides the test into four categories - test numbers and operations, geometry and measurements, algebra and functions, and statistics and probability - and further breaks these categories into understandable chapters so that the complexity of both the subject and the test is reduced. Ten chapters provide step-by-step instructions and simplify concepts by utilizing illustrated sample problems as well as comprehensive drills at the end of each lecture.

Linda M. Teel, East Carolina University, Greenville, NC

I was really impressed! It gave me insight of an easier way to look at a problem… one of the best study skills books I've come across.

Amneris Gonzalez, Instructional Support Services Secondary Math, LAUSD

Amy Lucas has an engaging, clear, conversational style… Explanations of difficult math concepts, especially the process for determining the different "plug-ins" used are quite excellent…. Step-by-step explanations are clear and concise.

Home Educating Family, Math section review

The tutorial on the essay section covers the important points in a clear and concise manner… the technique she employs will produce results… it works.

Nick Garrison, private tutor, Greenwich, CT

Lucas clearly states that she is coaching viewers on test-taking strategies, not teaching them general rules of grammar. Her examples are lucid and the suggestions consistent with those offered by other standardized test guides… VERDICT: Watching these DVDs and completing the workbook will not make the viewer a great writer, but the processes will help a student gain confidence in the writing skills required by the SAT.

Library Journal, Writing section review

I like the informal approach that the author uses to get on the student's level about the topics. Identifying each example and exercise with its relative difficulty also gives students a better sense of what to expect on the exam… well done; the portions where Amy gives her explanations with graphics are well thought out and produced…

David Hammett, Math Department Chair, Oakwood School, Los Angeles, CA

…she has an expressive style that should hold the attention of students. All incorrect responses are explained clearly. There are easy-to-follow tips for writing the essay as well as sample essays that received high and low scores.

Ellen Frank, School Library Media Specialist, Jamaica High School, Jamaica, NY

Students looking for a comprehensive review of the Writing portion of the SAT will find this program very helpful, especially since they can select specific areas in which they need improvement.

School Library Journal, Writing section review

This will really help the students get stronger in SAT. Great product for the students.

Sam G - parent, Houston

Introduction

Hi, I'm Amy Lucas and I'm going to be walking you through SAT Math, hopefully making it a bit more manageable, predictable, and a lot less scary. I will be addressing the math concepts tested, basic standardized test-taking tips, what types of problems to expect, and how to solve the different question types.

Before we delve into the SAT test, let's talk about why colleges require this test (as if years of school, homework, tests, and your subsequent GPA weren't enough!). Believe it or not, colleges aren't requiring the SAT or the ACT to cause additional anxiety or to make it that much harder to gain admittance. The SAT is designed as a fair measure of a student's readiness for college. It purports to test concepts learned and skills developed in school and to give an accurate assessment of a student's skills in logic, literacy, analysis, and writing.

Let me demystify the SAT for you. It does not test how smart you are. It does not test how well you will do in college. It doesn't really even test what you have learned in school; at least not in the same way you learned it. It's simply a test that tests how well you do on the SAT. Need more proof? The SAT used to stand for Scholastic Aptitude Test, and at one point it stood for Scholastic Assessment Test. Now it's just the SAT Reasoning Test. It doesn't stand for anything.

Some students are naturally good at taking standardized tests, but the majority of students aren't. You know that friend of yours that scored a 2100? Chances are she started out with an 1800 and got tutored, A LOT. While some students DO have the innate ability to problem solve, strategize, uncover tricks, not make careless mistakes, and remember concepts they learned in the second grade, the rest of us need to LEARN these skills. Trust me, you CAN get good at this test, and no one needs to know that you weren't naturally a standardized test prep genius.

Treat this program like a class you're taking in school. You are learning SAT Math. Some of it you might already know, but have forgotten. Other topics might be brand new to you. And remember, you don't get points for solving the problems the "right" way, so try not to be resistant to the techniques I introduce.

So who are the Mensa geniuses behind the SAT? The SAT is owned by the College Board, but administered by the Educational Testing Service. So I'm going to refer to the test makers as ETS. Many SAT experts will tell you that ETS is evil and out to get you. If that motivates you – use it! I personally think ETS could get a heck of a lot trickier. Either way, I'm here to help you master the game.

STRUCTURE

The math portion of the SAT is made up of 44 multiple-choice questions and 10 free-response 'grid-in' questions. The questions are divided into two 25-minute math sections and one 20-minute math section. The math sections can appear anywhere in the test. The essay will always be first, but after the essay they can either hit you with a math section, a critical reading section, or a writing section.

THE EXPERIMENTAL SECTION

You might end up with four math sections on the SAT. If that's the case, one of those four math sections is the experimental section. Leave it up to ETS to make you do an extra 25-minute math, writing, or critical reading section that doesn't count towards your score and makes the test a good 3 hours and 45 minutes long. ETS is using you as a guinea pig to test out future SAT questions.

There is no way to know if you're on an experimental section. It might just be the first math section you hit. All you know is that if you had three 25-minute math sections, it was one of the three, and if you had two 20-minute math sections, it was one of the two. This means, treat every section as if it counts, and don't let any section rattle you. Think of it this way: if you bomb the first test of the semester, it doesn't mean you can't get great scores on the other tests and still get an A in the class.

CONTENT

ETS tests numbers and operations, algebra and functions, geometry and measurement, data analysis, statistics, and probability. That's just arithmetic, algebra, geometry, and a little bit of algebra II.

Reference Box

$A = \pi r^2$
$C = 2\pi r$

$A = lw$

$A = \frac{1}{2} bh$

$V = lwh$

$V = \pi r^2 h$

$c^2 = a^2 + b^2$

Special Right Triangles

The number of degrees of arc in a circle is 360.

The sum of the measures in degrees of the angles of a triangle is 180.

You always get a reference box at the beginning of every math section. This reference box lists key formulas, such as area and circumference of a circle, and volume of a cylinder. If you forget a formula, or how many degrees are in a circle, or your 45-45-90 or 30-60-90 triangle rules, then check the reference box! You should know these formulas by test day, but there's no shame in looking.

Multiple Choice Questions

17. How many different integers made up of four nonzero digits can be formed if the tens digit must be 4, the ones digit cannot be 9, and digits may not be repeated?

(A) 84
(B) 294
(C) 448
(D) 648
(E) 2016

Answer: (B) 294

The SAT features standard multiple-choice questions. On multiple-choice questions you get ¼ of a point off for each wrong answer. Don't be afraid to take educated guesses, especially if you can eliminate answer choices, but don't be afraid to leave answers blank either. You don't get penalized for answers left blank. Chances are, if you only do questions 1-15 in each math section, and you miss no more than 2 or 3 of those questions, you'll score about a 600 on the math portion of the SAT.

If there is a figure or chart, it belongs to the question directly underneath it.

City	Number of Consecutive Days
A	7
B	3
C	5
D	8
E	6
F	7

16. The table above shows the number of consecutive days that each of five cities in Rocksford County lost power during a twelve–day period. If City C's power outage did not overlap with City F's power outage, which of the 12 days could be a day when only one city lost power?

(A) The 4th
(B) The 5th
(C) The 6th
(D) The 7th
(E) The 8th

Answer: (A) The 4th

Grid-Ins

I refer to the student-produced response questions as "grid-ins" because you bubble your answers into a grid. You don't get any points deducted for wrong answers on the grid-ins. You only get points added for right grid-in answers. So always take a guess on the grid-ins.

Grid-In Rules:

• Always write your answer in the boxes at the top. It'll help you avoid careless mistakes.

• Start gridding in at the left hand box. For instance, if your answer is 2, write 2 in the left hand column and then bubble in 2 underneath. Technically, you can bubble in 2 wherever you want, but for consistency's sake, start at the left.

• The exception is 0. Notice how the first column doesn't have a 0 to bubble in. So you should bubble in 0 in the second column.

- If you get a decimal, don't bother rounding. Just take whatever you see in the calculator, let's say 2.1289... and just grid in 2.12. That way you won't accidentally round wrong. If you were to just grid in 2.1, you would not get that point.

- You don't have to reduce fractions on grid-ins, unless of course they don't fit in the grid-in box. If I got 12/15 as my answer, I would either need to reduce it to 4/5 or change it into a decimal.

- You cannot grid in mixed numbers. If my answer was 2 and 1/3 it would be read as 21/3, so change it to an improper fraction.

- There are no negative answers on the grid-in section, so if you get a negative, something went wrong.

- Often, ETS will indicate "disregard the dollar sign" or "disregard the percent sign." If my answer was $3.15, I would grid in 3.15. If my answer was 23%, I would grid in 23, not .23, because .23 would indicate .23 %.

PROCESS OF ELIMINATION

Process of Elimination comes in handy on any standardized test. Often, the right answer choice is hidden among some very tempting wrong answers. The key is learning to spot and eliminate these wrong answers in order to increase your odds of choosing correctly. Every elimination increases your odds!

Examples of tempting answers to eliminate include partial answers. For instance, let's say ETS asks: "What is one-third of the value, in pounds, indicated on the scale's display?" I guarantee one of the answer choices will be the entire value indicated on the display, which is tempting to choose if you forget to reread what the question is asking for and neglect to divide the total amount into thirds.

Remember, SAT Math is arranged in order of difficulty. So if you are on an easy problem go ahead and pick that obvious answer, but if you are on a medium or difficult problem, beware of that obvious, easy answer! It is probably a trap that you should eliminate.

Statistically, if you can eliminate one answer choice it works to your advantage to take a guess. I would only take a guess if you have already left several other problems blank. Typically, I take a guess if I can eliminate 2 (definitely 3) answer choices.

CALCULATORS AND PENCILS

The good news is that you get to use a calculator. Start using the calculator you are going to use on the SAT so that you become familiar with it. Bring extra batteries the day of the test. Cell phones don't count as calculators!

Get used to using non-mechanical pencils. As of right now, ETS only allows number 2 pencils.

HOW TO USE THIS BOOK

I've structured this book so that you can skip around to different chapters if you already know your strengths and weaknesses. I encourage you to work through the whole book, but if you are a higher scoring student you might only need coaching on the trickier concepts.

Each lesson contains a lecture on the different topics and question types with sample problems illustrating each, and a comprehensive drill at the end. An answer key and explanation of the problems follow every comprehensive drill.

Don't be confused by the numbering of the questions. The math portion of the SAT is arranged in order of difficulty. I base my numbering system on a 20-question math section. So problem numbers 1-7 are easy, 8-15 medium, and 16-20 hard. If you are working through a number 5, you know it's easy, but if you are working through a number 18, you know it's a difficult problem. So if you're on the drill and you see a number 6, another number 6, and then a number 10, just know that the numbering is not off; I am simply indicating the question's level of difficulty.

So enough jabbering about the test and its structure! Let's figure out just what the SAT tests, and the clever ways ETS tries to trick us.

Chapter 1
Math Vocabulary

Here's a secret about SAT Math – it's simple! I know what you're thinking: *I can't seem to score above a 510. It seems pretty difficult to me*. The concepts tested are easy; the way ETS tests those concepts is what's tricky.

In some cases, the issue isn't the math at all – it's the reading comprehension. Students will misread the question, or misunderstand what the question is asking them to do. More often than not, it's a simple issue of vocabulary. So let's start by learning some basic SAT Math vocabulary. Get in the habit of underlining or circling these words whenever you see them in a question.

INTEGER

This is one of ETS's favorite words!

Some examples of integers are 2, 26, 3 . . .

But think outside the box; what about negative integers? –4, –42

What about 0? Yes, 0 is an integer too.

Definition: An integer is any number, positive, negative or zero, but no fractions and no decimals

---

NUMBER

I know you know what a number is, but in ETS speak the word "number" means, **"Don't forget those Fractions and Decimals!"**

Definition: All numbers, positive, negative and zero, *including* fractions and decimals

Examples: 2, $\frac{1}{2}$, –.113124...

If ETS uses the terminology **REAL NUMBER**, it simply means any number that isn't an imaginary number.

IMAGINARY NUMBER: the square root of a negative number.

Example: $\sqrt{-1} = i$

Imaginary numbers are not tested on the SAT – we only need to know them for the purpose of understanding the term 'real number.'

The words *integer* and *number* are often preceded by the words *even* and *odd*.

EVEN

Definition: Any integer divisible by 2

Examples: 2, 4, 6 ... but don't forget those negatives: −8, −10

What about 0? Is it even or odd? Let's look at it on a number line:

−2	−1	0	1	2	
Even	**Odd**	**EVEN**	**Odd**	**Even**	**0 is EVEN !**

ODD

Definition: Any integer not divisible by 2

Examples: −1, −33, 5, 9

Remember: there is no such thing as even or odd fractions and decimals.

Positive and *negative* often qualify the words *integer* and *number* as well. But, unlike with *even* and *odd*, fractions and decimals can be positive and negative. So when a question uses the terminology "positive or negative number" be sure to remember those fractions and decimals!

POSITIVE

Definition: All numbers greater than 0

Examples: 1, 1/3, .629, 100, .0001

NEGATIVE

Definition: All numbers less than 0

Examples: −2, −.5, −3/4

Is 0 positive or negative? NEITHER! *Remember:* **0 is neutral**. Let's review our rules of zero.

RULES OF ZERO

Zero is an integer

Zero is even

Zero is neutral (non-positive and non-negative)

Zero divided by any number is 0

Example: $\frac{0}{2} = 0$

Any number divided by zero is undefined

Example: $\frac{6}{0}$ = undefined

Positive, *negative*, and *zero* are easy enough, but what if ETS uses the phrases *non-negative*, *non-zero*, or *not positive*?

NON-NEGATIVE

Definition: a number that is either zero or positive

NON-ZERO

Definition: a number that is either positive or negative

NOT POSITIVE

Definition: a number that is either zero or negative

DIGIT

Definition: the integers 0 – 9

Examples: $0, 1, 2, 3, 4, 5, 6, 7, 8, 9$

So $6,435$ is a *four-digit number* made up of the digits $6, 4, 3,$ and 5.

DIGIT PLACES

Example: 4852.179
The place values are as follows:
   4 = the thousands digit
   8 = the hundreds digit
   5 = the tens digit
   2 = the units digit
   1 = the tenths digit
   7 = the hundredths digit
   9 = the thousandths digit

ETS loves to refer to the number in the ones place as the *units* digit. *Remember:* a digit is a single place holder and the hundreds digit is 8, not 800.

A word that often comes before *digit* is *distinct*.

DISTINCT

Definition: different

Example: *x* and *y* are distinct integers.

This simply means *x* and *y* cannot be the same integer.

Here are some no brainers:

SUM

Definition: the answer to an addition problem

Example: $57 + 6 = 63$

In the equation, 63 is the *sum* of 57 and 6.

DIFFERENCE

Definition: the answer to a subtraction problem

Example: $24 - 10 = 14$

In the equation, 14 is the *difference* between 24 and 10 (or "10 less than 24").

PRODUCT

Definition: the answer to a multiplication problem

Example: $3 \times 6 = 18$

In the equation, 18 is the *product* of 3 and 6.

QUOTIENT

Definition: the answer to a division problem

Example: $78 \div 3 = 26$

In the equation, 26 is the *quotient* when 78 is divided by 3.

Speaking of division, what does *divisible* mean?

DIVISIBLE

Definition: to divide into evenly

Example: 12 is *divisible* by 3 because it produces no remainder. ($12 \div 3 = 4$)

REMAINDER

Definition: the amount left over when a number doesn't divide into another number evenly

Example:

$$\begin{array}{r} 3\ R2 \\ 5\overline{)17} \\ -15 \\ \hline 2 \end{array}$$

When we divide 17 by 5 we get a *remainder* of 2. Long division is always a safe bet, but if long division isn't your thing, you can use your calculator.

Punch another example into your calculator now: $16,538 \div 3 = 5512.666667$
Is 6 your remainder? NO! Here are the steps to take to determine the remainder:

1. Subtract everything to the left of the decimal

   $5512.666667 - 5512 = .666667$

2. Multiply by the original *divisor* (the number you divide by)

   $.666667 \times 3 = 2$
   2 is your remainder!

The word *prime* is all over the SAT. There are a couple of ways ETS will test the concept of prime.

PRIME

Definition: A positive integer that is divisible only by 1 and itself

Examples: 5, 11, 17

Note that you only have to consider positive numbers when dealing with primes. So what's the smallest prime number? 3? 1? 2!

2 IS THE SMALLEST AND ONLY EVEN PRIME NUMBER!

Always keep in mind that 1 is NOT a prime number. Here's why: A prime number has to have two distinct factors, 1 and itself, and technically 1 only has one factor; 1 is divisible only by 1, so don't get tricked and count 1 as prime!

What about 0?

Remember how zero divided by any number is 0? That means that 0 has infinite factors. *Remember:* 0 AND 1 ARE NOT PRIME!

Factor and *multiple* are words that have been following us around since the third grade. ETS is counting on you to confuse them, SO DON'T!

FACTOR

Definition: a number that divides evenly into another (usually larger) number

Example: Factors of 18 are 1, 18, 2, 9, 6, 3

It's always helpful to list factors in pairs to make sure you don't miss any.

So: 1, 18 ($1 \times 18 = 18$)
2, 9 ($2 \times 9 = 18$)
3, 6 ($3 \times 6 = 18$)

Note that another name for *factor* is *divisor*.

Along with factors comes the terminology **PRIME FACTORIZATION**. Whenever ETS asks for the prime factors or the prime factorization of a number MAKE A FACTOR TREE!

Let's find the prime factorization of 252:

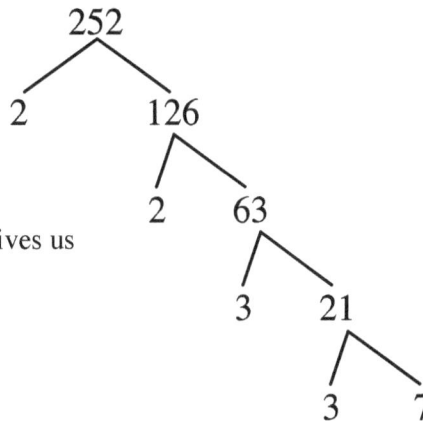

Breaking 252 down into its primes gives us

$2 \times 2 \times 3 \times 3 \times 7$

or

$252 = 2^2 \times 3^2 \times 7$

MULTIPLE

Definition: the numbers obtained by multiplying a given number by a series of integers

Example: Some multiples of 8 are **8** ($8 \times 1 = 8$), **16** ($8 \times 2 = 16$), **64** ($8 \times 8 = 64$), **–16** (don't forget negatives! $8 \times -2 = -16$), **0** ($8 \times 0 = 0$ - don't forget 0; zero is a multiple of everything!)

You don't have to worry about fractions and decimals when it comes to finding multiples.

What are the first five *consecutive positive multiples* of 5? ___ ___ ___ ___ ___

CONSECUTIVE

Definition: in a row from least to greatest

So the first five positive consecutive multiples of 5 are 5, 10, 15, 20, and 25.

_____

Factor and *multiple* aren't the only terms that people mix up. Students often don't read carefully enough and confuse "the square of" with "the square root of."

THE SQUARE OF: x^2

THE SQUARE ROOT OF: $\sqrt{x}$

THE CUBE OF: x^3

THE CUBE ROOT OF: $\sqrt[3]{x}$

The variables x and y will be all over the test, and I'll show you later the best way to deal with the algebra, but let's first discuss the difference between linear and quadratic equations.

LINEAR

Definition: refers to a straight line

Examples: $2x + 3y = 10$ or $2x + 3 = 7$. Both equations denote straight lines.

QUADRATIC

Definition: an equation with a squared variable that refers to a parabola

Examples: $x^2 + 7x + 12 = 0$ or $y = x^2 + 7x + 12$

The word *root* comes into play when dealing with *quadratics*.

ROOT

Definition: the solution to a quadratic equation (or where the graph of an equation crosses the *x*-axis)

Example: The roots of the quadratic equation $x^2 + 7x + 12 = 0$ are -4 and -3.

Explanation: To find the roots, first factor the quadratic equation.
$x^2 + 7x + 12 = 0$ factors to $(x+4)(x+3) = 0$.
So, set $x + 4 = 0$ and $x + 3 = 0$ then solve for x twice!

$x + 4 = 0$	$x + 3 = 0$
$x + 4 - 4 = 0 - 4$	$x + 3 - 3 = 0 - 3$
We subtract 4 from both sides, so that we can cancel out the 4 on the left $(4 - 4 = 0)$ and get x alone.	We subtract 3 from both sides, so that we can cancel out the 3 $(3 - 3 = 0)$ on the left and get x alone.
$x = -4$	$x = -3$

The *factors* are $(x + 4)$ and $(x + 3)$, but the *roots* are $-4, -3$

Let's discuss the difference between the symbols (,) and { , }

(,) **refers to an ordered pair** (x, y) **that stands for a point found on the coordinate grid.**

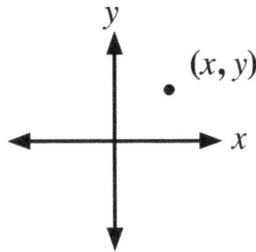

{ , } **refers to a set of numbers or elements.**

Example: The set of odd digits is: $\{1, 3, 5, 7, 9, \ldots\}$

Union and *intersection* are words that appear in problems that test the concept of *sets*.

UNION

Definition: a set made up of the combined elements (or numbers) of two or more sets

Union is represented by the symbol U

Example: $\{3, 7, 6, 8, 10\} \cup \{2, 3, 5, 6, 8, 11\} = \{2, 3, 5, 6, 7, 8, 10, 11\}$

Be careful: when the sets share a common number only list that number once!

INTERSECTION

Definition: a set made up of only elements that are in both of two other sets

Intersection is represented by the symbol $\cap$

Example: $\{3, 7, 6, 8, 10\} \cap \{2, 3, 5, 6, 8, 11\} = \{3, 6, 8\}$

The only other thing to know about sets is the *empty set.*

EMPTY SET

Definition: exactly what it sounds like — a set that contains no elements

Represented by { } or Ø.

Let's circle back to the (x, y) ordered pair. Occasionally the terms *domain* and *range* show up on the SAT.

DOMAIN

Definition: all x values used for a function

Example: $y = \dfrac{7}{x-2}$

The domain = all real numbers except 2. This is because $2-2=0$ and $\dfrac{7}{0}$ is undefined, so the thing to remember about domain is that the divisor can never equal 0. The only number that results in a divisor of 0 is 2, so the domain is all numbers except 2.

RANGE

Definition: all y values for a function

ETS generally gives you a specified domain, so throw in some x values to determine your y values.

Example: $f(x) = -3x + 6$.

The domain of the function above is $-5 \le x \le 5$. Throw in the endpoints -5 and 5 of the domain in order to determine your range.

$$f(x) = -3x + 6 \qquad\qquad f(x) = -3x + 6$$
$$= -3(-5) + 6 \qquad\qquad = -3(5) + 6$$
$$= 15 + 6 \qquad\qquad\qquad = -15 + 6$$
$$= 21 \qquad\qquad\qquad\quad = -9$$

So, the range is $-9 \le y \le 21$.

Absolute Value shows up on the trickier math questions (which become much simpler with a little technique and practice) but absolute value itself is easy enough.

ABSOLUTE VALUE

Definition: a number's distance from zero on the number line

Examples: $|6| = 6, |-6| = 6$

If $|x| = 6$, then $x = 6$ or -6

Lastly, we have some statistical terms:

MEAN

Definition: the average

$$\text{Average} = \frac{\text{Sum}}{\text{\# of things}}$$

Example: *Mean* of 3, 7, 8, and 2 is 5

$$= \frac{3 + 7 + 8 + 2}{4}$$
$$= \frac{20}{4}$$
$$= 5$$

MEDIAN

Definition: the middle value of a set when the numbers are arranged from least to greatest

Examples: {3, 2, 9, 8, 3}
First, arrange the numbers in the set from least to greatest: {2, 3, 3, 8, 9}
There are five numbers in the set, so the middle number is the third number in the set, and the third number is 3. So, *median = 3*

{5, 3, 10, 7}
Again, arrange the numbers in order from least to greatest: {3, 5, 7, 10}
Notice that there is an even number of elements in the set, so there is no single number in the exact middle. **When there is an even number of elements, find the average of the two middle numbers.** So, *median = 6*, because $\frac{5+7}{2} = \frac{12}{2} = 6$

MODE

Definition: the number that occurs most frequently in a set

Examples: {2, 2, 3, 5, 4}
The mode of this set is 2, because 2 occurs more frequently (twice) than any of the other members of the set (3, 5, 4), all of which occur only once.

{2, 2, 3, 3, 4}
Both 2 and 3 occur more frequently than 4, but neither 2 nor 3 occurs more than the other, so this set has **two modes**, 2 and 3.

{2, 2, 3, 3}
Does 2 occur more than 3? Does 3 occur more than 2? Don't let them trick you. No member of this set occurs more often than any other member of the set, so this set has NO MODE!

Remember: each set has only one mean and one median but can have multiple modes.

That pretty much covers each of ETS's favorite words. It's amazing how easy the words are to miss! Underline them, circle them, and pay attention! Often ETS will underline or capitalize these words and students STILL gloss over them. Beware! Now, let's put all these terms together and see how you do on the comprehensive drill that follows.

Math Vocabulary Drill

2. $10^3 - 100$ is divisible by all of the following *EXCEPT*

(A) 15
(B) 45
(C) 50
(D) 60
(E) 170

2. If G is the set of non-negative integers, M is the set of even integers, and O is the set of integers less than 4, which of the following integers is a member of all three sets?

(A) –2
(B) 0
(C) 1
(D) 4
(E) 6

2. The sum of $4x$ and the cube root of y is equal to the cube of the sum of x and y.

Which of the following expresses the statement above?

(A) $\sqrt[3]{4x} + y = x^2 + y^3$
(B) $\sqrt[3]{4x} + \sqrt[3]{y} = (x + y)^3$
(C) $4x + \sqrt[3]{y} = (x + y)^3$
(D) $4x + \sqrt[3]{y} = x^3 + y^3$
(E) $4x + y^3 = \sqrt[3]{x + y}$

7. How many two-digit integers between 17 and 101, inclusive, have the tens digit equal to 3, 6, or 9 and the units digit equal to 4, 5, or 7?

(A) Three
(B) Four
(C) Six
(D) Nine
(E) Ten

9. If 2.874 is rounded to the nearest tenth, the result is how much less than if 2.874 is rounded to the nearest whole number?

10. Set Q contains the positive multiples of 4 that are less than 60 and set R contains the positive multiples of 6 that are less than 60. How many elements are in set Q∩R?

(A) None
(B) One
(C) Two
(D) Three
(E) Four

11. The sum of the consecutive integers from −32 to y is 138. What is the value of y?

(A) 33
(B) 35
(C) 36
(D) 72
(E) 132

$$2, 3, 5, 8, 9, 11, 12, 15$$

12. The integer x is to be added to the list of numbers above. Which of the following could be the median of the new list?

I. 8
II. 8 1/2
III. 9

(A) I only
(B) II only
(C) II and III only
(D) I and III only
(E) I, II, and III

13. A survey of pets in city A showed that there was an average of 1.3 pets per person and an average of 2.5 children per person. Based on the results of the survey, if 52,000 pets inhabit city A, which of the following is the best approximation for the total number of children in city A?

(A) 40,000
(B) 52,550
(C) 100,000
(D) 123,000
(E) 134,000

13. If $n = a \cdot b \cdot c$, and a, b, c are three distinct odd prime numbers, how many positive factors does n have?

$$7, 4, 3, 6, 3, 4, 3, n, 4, 3, 6$$

13. For the numbers listed above, the median is 4 and the only mode is 3. Which of the following numbers could not be the value of n?

(A) 3
(B) 4
(C) 6
(D) 7
(E) 8

$$|x - 4| = \frac{1}{5}$$

14. What is the least value of x that makes the equation true?

14. The sum of thirteen different integers is zero. What is the least number of these integers that must be negative?

(A) Six
(B) Three
(C) Two
(D) One
(E) None

14. If the difference between k and n is odd, and the value of $(k + n)^2 + r + k$ is even, and $k, n,$ and r are non-zero integers, then which of the following must be true?

(A) k is even
(B) n is even
(C) kn is odd
(D) If r is even, then kn is odd
(E) If r is even, then k is odd

15. What is the product of the greatest prime number that is less than 30 and the smallest prime number that is greater than 30?

18. Nine times the square of m is equal to the square of n. If n is 2 more than three times m, then what is the value of n?

(A) -3

(B) -1

(C) $-\frac{1}{3}$

(D) $\frac{1}{3}$

(E) 1

Answers and Explanations

Answer Key:

2. (E)	**9.** .1	**13.** (C)	**14.** (D)
2. (B)	**10.** (E)	**13.** 8	**14.** (E)
2. (C)	**11.** (C)	**13.** (B)	**15.** 899
7. (D)	**12.** (D)	**14.** 19/5	**18.** (E)

2. $10^3 - 100$ is divisible by all of the following *EXCEPT*

(A) 15
(B) 45
(C) 50
(D) 60
(E) 170

Explanation:
$10^3 - 100 = 1000 - 100 = 900$

(A) $\dfrac{900}{15} = 60$

(B) $\dfrac{900}{45} = 20$

(C) $\dfrac{900}{50} = 18$

(D) $\dfrac{900}{60} = 15$

(E) $\dfrac{900}{170} = 5.29411764...$

Of the answer choices, only 170 doesn't evenly divide 900.

---

2. If G is the set of non-negative integers, M is the set of even integers, and O is the set of integers less than 4, which of the following integers is a member of all three sets?

(A) –2
(B) 0
(C) 1
(D) 4
(E) 6

Explanation:
Write out your sets and the answer is easy to spot:
Set G: {0, 1, 2, 3, 4, 5, 6, 7,...}
Set M: {... –2, 0, 2, 4, 6, 8,...}
Set O: {...–3, –2, –1, 0, 1, 2, 3}

Go through the answer choices:
    (A) –2
–2 couldn't be in G, because all the members of G are non-negative.
    (B) 0
0 is in all three sets; 0 is non-negative, even, and an integer less than 4.
    (C) 1
1 couldn't be in M, because it is not even.

(D) 4

4 could not be in O, because it is not less than 4.

(E) 6

6 could not be in O, because it is not less than 4.

The only number in the answer choices that could be in all three sets is 0.

---

2. The sum of $4x$ and the cube root of y is equal to the cube of the sum of x and y.

Which of the following expresses the statement above?

(A) $\sqrt[3]{4x} + y = x^2 + y^3$
(B) $\sqrt[3]{4x} + \sqrt[3]{y} = (x + y)^3$
(C) $4x + \sqrt[3]{y} = (x + y)^3$
(D) $4x + \sqrt[3]{y} = x^3 + y^3$
(E) $4x + y^3 = \sqrt[3]{x + y}$

Explanation:

Translate each portion of the statement into mathematical symbols:

The sum of $4x$ and the cube root of y is equal to the cube of the sum of x and y

$$4x + \sqrt[3]{y} = (x + y)^3$$

---

7. How many two-digit integers between 17 and 101, inclusive, have the tens digit equal to 3, 6, or 9 and the units digit equal to 4, 5, or 7?

(A) Three
(B) Four
(C) Six
(D) Nine
(E) Ten

Explanation:

Write out the possible combinations:

34 1
35 2
37 3
64 4
65 5
67 6
94 7
95 8
97 9

There are nine possible combinations.

---

9. If 2.874 is rounded to the nearest tenth, the result is how much less than if 2.874 is rounded to the nearest whole number?

Answer: .1

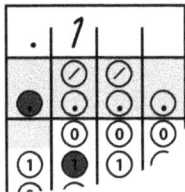

Explanation:
2.874 rounded to the nearest tenth = 2.9
2.874 rounded to the nearest whole number = 3
"how much less than" means to subtract: 3 − 2.9 = .1

10. Set Q contains the positive multiples of 4 that are less than 60 and set R contains the positive multiples of 6 that are less than 60. How many elements are in set Q∩R?

(A) None
(B) One
(C) Two
(D) Three
(E) Four

Explanation:
Set Q: {4, 8, 12, 16, 20, 24, 32, 36, 40, 44, 48, 52, 56}
Set R: {6, 12, 18, 24, 30, 36, 42, 48, 54}

Q∩R : {12, 24, 36, 48}
The set has four members.

11. The sum of the consecutive integers from −32 to y is 138. What is the value of y?

(A) 33
(B) 35
(C) 36
(D) 72
(E) 132

Explanation:
If we were to add all the integers from −32 to 0, we would get a certain negative number. If we were to add up all the integers from 0 to 32, we would get the same number only positive instead of negative, so the sum of the consecutive integers from −32 to 32 = 0. But the sum we're looking for is 138, not 0, so we need to go beyond 32. Keep adding consecutive integers after 32 until you get 138:

33 = 33
33 + 34 = 67
33 + 34 + 35 = 102
33 + 34 + 35 + 36 = 138
So, y = 36

$$2, 3, 5, 8, 9, 11, 12, 15$$

12. The integer x is to be added to the list of numbers above. Which of the following could be the median of the new list?

I. 8
II. 8½
III. 9

(A) I only
(B) II only
(C) II and III only
(D) I and III only
(E) I, II, and III

Explanation:
Let $x = 8$
So the new list is $2, 3, 5, 8, 8, 9, 11, 12, 15$
There is an odd number of numbers in the list and the middle number when the set is arranged from least to greatest is 8, so 8 is the new median, and I is a possibility.

Let $x = 9$
So the new list is $2, 3, 5, 8, 9, 9, 11, 12, 15$
9 is the new median, and III is a possibility. Because the list will always contain an odd number of integers, the median will always be the middle number, and the middle number has to be an integer because of the given restriction "the integer x," so there is no way to get 8½ as the median.

---

13. A survey of pets in city A showed that there was an average of 1.3 pets per person and an average of 2.5 children per person. Based on the results of the survey, if 52,000 pets inhabit city A, which of the following is the best approximation for the total number of children in city A?

(A) 40,000
(B) 52,550
(C) 100,000
(D) 123,000
(E) 134,000

Explanation:
This is a proportion problem:
$$\frac{\text{Pets}}{\text{Children}} = \frac{1.3}{2.5} = \frac{52,000}{x}$$

Cross-multiply to solve:
$$\frac{1.3}{2.5} \diagdown \frac{52,000}{x}$$

$(2.5)(52,000) = 1.3x$

$130,000 = 1.3x$

$\dfrac{130,000}{1.3} = \dfrac{1.3x}{1.3}$ ⟶ 1.3 divided by 1.3 is just 1, so the 1.3 cancels out and the x is left alone.

$x = 100,000$

13. If $n = a \bullet b \bullet c$, and a, b, c are three distinct odd prime numbers, how many positive factors does n have?

Answer: 8

Explanation:
Let $a = 3, b = 5, c = 7$
$n = a \bullet b \bullet c$
$n = 3 \bullet 5 \bullet 7$
$3 \bullet 5 \bullet 7 = 105$
$n = 105$

Positive Factors of 105: $1, 105$ ($1 \times 105 = 105$)
$3, 35$ ($3 \times 35 = 105$)
$5, 21$ ($5 \times 21 = 105$)
$7, 15$ ($7 \times 15 = 105$)

The factors of 105 are $1, 105, 3, 35, 5, 21, 7$, and 15, so 105 has 8 factors.

$$7, 4, 3, 6, 3, 4, 3, n, 4, 3, 6$$

13. For the numbers listed above, the median is 4 and the only mode is 3. Which of the following numbers could not be the value of n?

(A) 3
(B) 4
(C) 6
(D) 7
(E) 8

Explanation:
The easiest way to solve this problem is to plug in the answer choices and find the one that doesn't match the restrictions.
(A) Let $n = 3$
$7, 4, 3, 6, 3, 4, 3, 3, 4, 3, 6$
Reorder the set consecutively: $3, 3, 3, 3, 3, 4, 4, 4, 6, 6, 7$
In this list, 4 is the median and 3 is the only mode, so 3 is a possible value for n.

(B) Let $n = 4$
$7, 4, 3, 6, 3, 4, 3, 4, 4, 3, 6$
Reorder the set consecutively: $3, 3, 3, 3, 4, 4, 4, 4, 6, 6, 7$
The list contains four 4s and four 3s, violating the restriction that 3 is the only mode.

$$|x - 4| = \tfrac{1}{5}$$

14. What is the least value of x that makes the
equation true?

Answer: $\dfrac{19}{5}$

Explanation:

$x - 4 = \tfrac{1}{5}$ or $x - 4 = -\tfrac{1}{5}$

$x - 4 + 4 = \tfrac{1}{5} + 4$ $x - 4 + 4 = -\tfrac{1}{5} + 4$

$x = 4\tfrac{1}{5}$ $x = 3\tfrac{4}{5}$

x could be either $3\tfrac{4}{5}$ or $4\tfrac{1}{5}$. The least is $3\tfrac{4}{5}$, but you can't grid in a mixed
fraction, so convert to an improper fraction before gridding in.

$3\tfrac{4}{5} = \dfrac{19}{5}$

---

14. The sum of thirteen different integers is zero.
What is the least number of these integers that
must be negative?

(A) Six
(B) Three
(C) Two
(D) One
(E) None

Explanation:
Plug in numbers to make this abstract concept concrete. Since the question asks
for the least number of the integers that has to be negative, keep as many of the
integers positive for as long as possible until you have to make one negative. Add
together the integers 1 through 12.

$$1 + 2 + 3 + 4 + 5 + 6 + 7 + 8 + 9 + 10 + 11 + 12 = 78$$

The sum is 78, so the thirteenth integer could be -78, which would bring
the total sum to 0.

Therefore, the only negative integer in the set of 13 is -78.

---

14. If the difference between k and n is odd, and the value of $(k + n)^2 + r + k$ is even, and $k, n,$ and r are non-zero integers, then which of the following must be true?

(A) k is even
(B) n is even
(C) kn is odd
(D) If r is even, then kn is odd
(E) If r is even, then k is odd

Explanation:
Who the heck knows when all they give us are variables!
Throw in numbers to make the problem a simple arithmetic problem.

Let $k = 4, n = 3, r = 1$
$4 - 3 = 1$ (matches restriction, because 1 is odd)
$(4 + 3)^2 + 1 + 4 = 54$ (matches restriction, because 54 is even)

Now let's plug our numbers into the answer choices. This eliminates (B), because 3 is not even. It also eliminates (C), because $4 \times 3 = 12$, which is not odd.

Let's make k an odd number and n an even number this time.
Let $k = 5, n = 2, r = 2$
$5 - 2 = 3$ (matches restriction)
$(5 + 2)^2 + 2 + 5 = 56$ (matches restriction)

This eliminates (A), because 5 is not even. It also eliminates (D), because 2 is even, but $5 \times 2 = 10$, which is not odd. So the only answer choice that has not been eliminated is (E).

---

15. What is the product of the greatest prime number that is less than 30 and the smallest prime number that is greater than 30?

Answer: 899

Explanation:
Set of primes:
$\{2, 3, 5, 7, 11, 13, 17, 19, 23, 29, 31, 37, ...\}$

The greatest prime number that is less than 30 is 29.
The smallest prime number that is greater than 30 is 31.

$29 \cdot 31 = 899$

18. Nine times the square of m is equal to the square of n. If n is 2 more than three times m, then what is the value of n?

(A) -3
(B) -1
(C) $-\frac{1}{3}$
(D) $\frac{1}{3}$
(E) 1

Explanation:

Translate "Nine times the square of m is equal to the square of n" into a mathematical expression: $9m^2 = n^2$

Translate "n is 2 more than three times m" into a mathematical expression: $n = 3m + 2$

If we plug in the stated value for n into the equation $9m^2 = n^2$ we get:

$$9m^2 = (3m + 2)^2$$

Solve for m:

$$9m^2 = (3m + 2)(3m + 2)$$
$$9m^2 = 9m^2 + 6m + 6m + 4$$
$$9m^2 = 9m^2 + 12m + 4$$
$$9m^2 - 9m^2 = 9m^2 + 12m + 4 - 9m^2$$
$$0 = 12m + 4$$
$$0 - 4 = 12m + 4 - 4$$
$$-4 = 12m$$
$$\frac{-4}{12} = \frac{12m}{12}$$
$$m = -\frac{1}{3}$$

Now plug the determined value for m into the equation and solve for n

$$n = 3\left(-\frac{1}{3}\right) + 2$$
$$n = \left(-\frac{3}{3}\right) + 2$$
$$n = -1 + 2$$
$$n = 1$$

There is an even easier way to solve this problem that lets us avoid the algebra. Notice how there are numbers in the answer choices and the question asks for a single, specific thing: "what is n?" One of the answer choices has to be n, so why not work backwards and plug in our answer choices for n? Typically we start with answer choice (C), because we will have a good idea which number to try next, bigger or smaller, if (C) doesn't work. Since we already know the correct answer is E, let's just plug in 1 to demonstrate this technique.

$$n = 1$$
$$1 = 3m + 2$$
$$1 - 2 = 3m + 2 - 2$$
$$-1 = 3m$$
$$\frac{-1}{3} = \frac{3m}{3}$$
$$m = -\frac{1}{3}$$

Now test it with the other equation:

$$9\left(-\frac{1}{3}\right)^2 = 1^2$$
$$9\left(\frac{1}{9}\right) = 1$$
$$\frac{9}{9} = 1$$
$$1 = 1 \checkmark$$

Chapter 2
Basic Plugging In

The SAT is full of algebra, and while it's handy to know how to work with variables, our goal is to turn every single algebra problem we can into an arithmetic problem. Why? Because we don't think in terms of algebra. We don't go out to dinner and say, you ate like a pig so you owe $2x$ and I ate like a bird so I owe $\frac{1}{2}x$. We think in terms of numbers.

So repeat this mantra: **When in doubt, plug in!**

There are three types of plugging in problems. Let's talk about the first – *The Basic Plug In*.

Because there are three different types of plug in problems scattered throughout the SAT it is important to know how to identify and differentiate between them. So let's take a look at the following Basic Plug In problems. What do all of the answer choices have in common? VARIABLES.

Variables in answer choices means plug in!

Let's work through this simple algebra problem as a basic plug in, just to demonstrate the concept.

4. Vicky has cut hair 6 years less than twice as long as Jane has. If Jane has cut hair for m years, which of the following expressions represents the number of years that Vicky has cut hair?

    (A) $m - 6$
    (B) $m + 6$
    (C) $2m + 6$
    (D) $2m - 6$
    (E) $6 - 2m$

Explanation:
We have two unknowns in the question:
   1) the amount of time that Vicky has cut hair
   2) the amount of time that Jane has cut hair

These two unknowns are obviously dependent on each other. Start plugging in for the variable or unknown that gets the most action. In this case it's the number of years Jane has cut hair, so let's throw in a simple and easy number and set $m = 4$. Twice as long as Jane means 2 times 4 or 8, and 6 less than that translates to $8 - 6$, which means that Vicky has been cutting hair for 2 years.

The question asks for how long Vicky has cut hair, and we always want to put a box around whatever the question is asking for. So write down $v = 2$ and put a box around it. Once we have our boxed answer, it's time to go to the answer choices. We said that $m = 4$ so throw 4 into the answer choices to see which one gives us 2.

(A) $m - 6$
$4 - 6 = -2$

(D) $2m - 6$
$2(4) - 6 = 8 - 6 = 2$

(B) $m + 6$
$4 + 6 = 10$

(E) $6 - 2m$
$6 - 2(4) = 6 - 8 = -2$

(C) $2m + 6$
$2(4) + 6 = 8 + 6 = 14$

Answer: (D) $2m - 6$

Let's try another one:

19. An art gallery consultant collects a commission of w percent of the selling price of a piece of art. Of the following expressions, which represents the commission, in dollars, on 3 pieces of art that sold for $8,000 each?

(A) $240w$

(B) $4000w$

(C) $16000w$

(D) $\dfrac{8,000}{100 + w}$

(E) $\dfrac{16,000 + w}{100}$

Explanation:

Notice those variables in the answer choices and note that it's a basic plug in! Let's plug in for w. w stands for a percent and when dealing with percent problems it is always easiest to plug in 100, or 10.

So let's make $w = 10$.

If a piece of art sells for $8000, then let's figure out what ten percent of 8000 is:

$$\frac{10}{100} \times 8000 = \frac{80000}{100} = 800$$

And there are 3 pieces of art, so $800 \times 3 = \$2400$.

The question is asking for the commission in dollars on three pieces of art so put a box around 2400.

$$\boxed{2400}$$

Plug in 10 for w in the answer choices:

(A) $240w$
$240(10) = 2400$

(D) $\dfrac{8,000}{100 + w}$
$\dfrac{8,000}{100 + 10} = \dfrac{8,000}{110} = 72.\overline{72}$

(B) $4,000w$
$4,000(10) = 40,000$

(E) $\dfrac{16,000 + w}{100}$
$\dfrac{16,000 + 10}{100} = \dfrac{16,010}{100} = 160.1$

(C) $16,000w$
$16,000(10) = 160,000$

Answer: (A) $240w$

We see that our first choice, (A), gives us 2400. But be careful: we can't just stop at (A). *Because we are throwing in arbitrary numbers, there is a chance that we may get more than one answer choice working, so we have to try them all.*

Let's see this concept in action:

8. If $p < r$, and p, r, and m are positive integers, how much greater is the sum of m and r than the difference of m and p?

(A) $m - r$
(B) $2m - p$
(C) $r - p$
(D) $r + p$
(E) $2m - r - p$

Explanation:
Plug in $p = 2, r = 3$, and $m = 5$
$2 < 3$ (matches restriction)
$m + r = 5 + 3 = 8$
$m - p = 5 - 2 = 3$
$\quad\quad 8 - 3 = 5$

$\boxed{5}$ "How much greater is the sum (8) than the difference (3)?"

(A) $m - r$
$\quad 5 - 3 = 2$

(B) $2m - p$
$\quad 2(5) - 2 = 8$

(C) $r - p$
$\quad 3 - 2 = 1$

(D) $r + p$
$\quad 3 + 2 = 5$

(E) $2m - r - p$
$\quad 2(5) - 3 - 2 = 5$

Both (D) and (E) match, so pick new numbers.

This time let's make r bigger than both p and m. Plug in $p = 3$, $r = 8$, and $m = 5$
$3 < 8$ (matches restriction)
$m + r = 5 + 8 = 13$
$m - p = 5 - 3 = 2$ $\boxed{11}$
$\quad\quad 13 - 2 = 11$

We only need to try (D) and (E), because we've eliminated the others:
(D) $r + p$
$\quad 8 + 3 = 11$

(E) $2m - r - p$
$\quad 2(5) - 8 - 3 =$
$\quad 10 - 8 - 3 =$
$\quad 2 - 3 = -1$

This time only (D) gives us the match of 11.
Make sure you try all the answer choices on a Basic Plug In.

Answer: (D) $r + p$

BASIC PLUG IN TIPS

- Plug in simple numbers such as 2, 3, and 5.
- Remember, no matter what number you plug in, if the arithmetic gets ugly, just go back and throw in another number!
- For this type of plug in you want to avoid using 0 or 1. Also, try not to throw in numbers that are in the answer choices. This will help us avoid getting more than one answer working.
- There are better numbers than others to use in conversion type problems. If the question says something like "z cents" throw in 100 for z. 100 cents is 1 dollar and believe me, somewhere in the question you'll have to convert those cents into dollars. Likewise, if they say "x inches" throw in 12, because 12 inches converts to 1 foot, "y minutes" throw in 60, and of course for percent problems throw in 100 or 10!

Let's try a few more:

18. If $k = n - h - j$, what is the average (arithmetic mean) of k, h, and j, in terms of n?

(A) $\dfrac{n+1}{3}$

(B) $\dfrac{3n}{4}$

(C) $3n$

(D) $\dfrac{n}{3}$

(E) $\dfrac{2n}{3}$

Explanation:
Notice those variables in the answer choices! And notice that expression "in terms of n." Put a line through it. It means nothing to us because we are turning this algebra problem into an arithmetic problem, so cross out the confusion! Let's plug in.

$h = 2,\ j = 3,\ n = 4$
$k = n - h - j$
$k = 4 - 2 - 3$
$k = 2 - 3$
$k = -1$

Average is the sum of k and h and j divided by the number of numbers, which is 3 so,

$$\frac{-1+2+3}{3} = \frac{1+3}{3} = \frac{4}{3} \qquad \boxed{\dfrac{4}{3}}$$

And then plug our n, which is 4, into the answer choices to find out the correct answer is (D).

(A) $\dfrac{n+1}{3}$ (D) $\dfrac{n}{3}$

$\dfrac{5+1}{3} = \dfrac{6}{3} = 2$ $\dfrac{4}{3}$

(B) $\dfrac{3n}{4}$ (E) $\dfrac{2n}{3}$

$\dfrac{3(4)}{4} = \dfrac{12}{4} = 3$ $\dfrac{2(4)}{3} = \dfrac{8}{3} = 2\dfrac{2}{3}$

(C) $3n$

$3(4) = 12$

Answer: (D) $\dfrac{n}{3}$

See that wasn't hard! And that was a number 18, one of the supposedly most difficult problems on the test.

Let's do one more:

13. If $p = r^2$ for any positive integer r, and if
$p^3 + p = q$, what is q in terms of r?

(A) $r^3 + r$
(B) $r^4 + r^2$
(C) $r^6 + r^2$
(D) $r^5 + r^3$
(E) r^4

Explanation:
Remember we get to cross out "in terms of r" so the question is simply asking what is q?
Let's make $r = 2$
$p = r^2$
$2^2 = 4$, so $p = 4$
$p^3 + p = q$
$4^3 + 4 = q$
$64 + 4 = q$
$q = 68$

68

Plug in 2 for r into each answer choice to show that (C) gives us 68:

(A) $r^3 + r$
$2^3 + 2$
$8 + 2 = 10$

(D) $r^5 + r^3$
$2^5 + 2^3$
$32 + 8 = 40$

(B) $r^4 + r^2$
$2^4 + 2^2$
$16 + 4 = 20$

(E) r^4
$2^4 = 16$

(C) $r^6 + r^2$
$2^6 + 2^2$
$64 + 4 = 68$

Answer: (C) $r^6 + r^2$

You can also use Basic Plug In for geometry problems. Just look for those variables in the answer choices. Let's do one together:

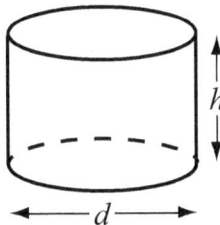

11. The right circular cylinder above has a volume of 32π and a height of h. If d represents the diameter, which of the following expressions represents the volume of the smallest rectangular box that completely contains the cylinder?

(A) dh
(B) d^2h^2
(C) $(d + h)^2$
(D) dh^2
(E) d^2h

Explanation:
They tell us that the volume of the cylinder is equal to 32π, so set the formula for volume of a right circular cylinder, $\pi r^2 h$, equal to 32π. Make sure to plug in for the height. Let's say $h = 8$:

$$h = 8$$
$$\pi r^2 h = \text{volume}$$
$$\pi r^2 8 = 32\pi$$
$$\frac{\pi r^2 8}{8\pi} = \frac{32\pi}{8\pi}$$
$$\frac{\pi r^2 8}{8\pi} = \frac{32\pi}{8\pi}$$ → The πs cancel out!
$$r^2 = 4$$
$$\sqrt{r^2} = \sqrt{4}$$
$$r = 2$$

So, d = 4

Now plug in for the diameter.
Draw in a rectangular solid surrounding the cylinder completely:

The volume of a rectangular solid is $l \times w \times h$.
The diameter is equal to the length and the width.
So solve for volume: $4 \times 4 \times 8 = 128$

128

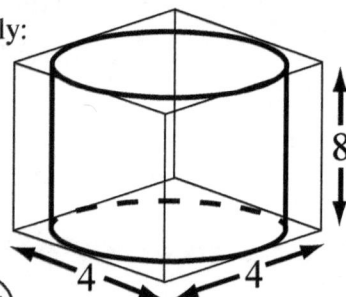

Box it! And plug in 4 for d and 8 for h into the answer choices:

(A) dh
    $4 \times 8 = 32$

(B) $d^2 h^2$
    $4^2 \times 8^2 =$
    $16 \times 64 = 1024$

(C) $(d+h)^2$
    $(4+8)^2 =$
        $(12)^2 = 144$

(D) dh^2
    $4 \times 8^2 =$
    $4 \times 64 = 256$

(E) $d^2 h$
    $4^2 \times 8 =$
    $16 \times 8 = 128$

Answer: (E) d^2h

Time to try some on your own!

Basic Plug In Drill

2. If x is an even integer, and y is an odd integer, which of the following is an even integer?

(A) $3x + y$
(B) $(x + y)^2$
(C) $2y + x^2$
(D) $2x + 3y$
(E) $3(x + y)$

6. If the average of a and $a + 4$ is b, and if the median of $a, a + 2, b$, and $b - 2$ is c, what is the average of a and b, in terms of c?

(A) 2
(B) c
(C) $\frac{c}{2}$
(D) $c + \frac{3}{2}$
(E) $c + 2$

6. Bob is three times as old as Jenny, who is five times as old as Lou. If Bob is x years old, then in ten years how old will Lou be in terms of x?

(A) $\frac{x}{5 + 10}$
(B) $\frac{x + 2}{2}$
(C) $\frac{x + 150}{15}$
(D) $\frac{3x}{5 + 10}$
(E) $\frac{3x}{3 + 10}$

7. Each of the following is equivalent to $\dfrac{x}{xy(xy^2 + z)}$ *EXCEPT*

(A) $\dfrac{1}{y(z + xy^2)}$

(B) $\dfrac{x}{x^2y^3 + xyz}$

(C) $\dfrac{1}{xy^3 + yz}$

(D) $\dfrac{1}{y^3 + yz}$

(E) $\dfrac{1}{y(xy^2 + z)}$

12. If y is a positive integer and $3^y + 3^{(y+1)} = z$, what is $3^{(y+1)}$ in terms of z?

(A) z^2
(B) $3z + 1$
(C) $\dfrac{3z}{4}$
(D) $\dfrac{(3-z)}{2}$
(E) $\dfrac{2z}{3}$

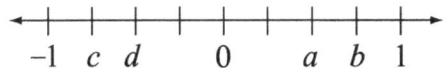

13. The letters c, d, a, and b represent numbers on the number line as shown above. Which of the following expressions has the least value?

(A) $c + b$
(B) $c + a$
(C) $d + a$
(D) $c - d$
(E) $b - a$

15. If $2n + 3m = m$, then in terms of n, which of the following must equal $4n + 4m$?

(A) 0
(B) 8
(C) $4n$
(D) $8n$
(E) $6n - 6$

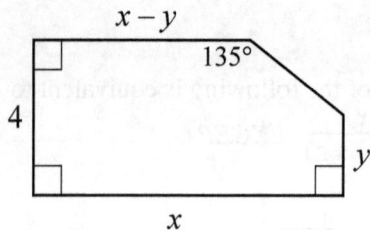

18. Which of the following represents the area of the figure above?

(A) $4x + y\sqrt{3}$

(B) $\dfrac{7x}{2} + y\sqrt{3}$

(C) $6x^2 + y\sqrt{3}$

(D) $\dfrac{7x}{2} + y^2$

(E) $4x - 2y + \dfrac{y^2}{2}$

20. A plumber charges y dollars for the first hour of work and charges for any additional time at the rate of z dollars per minute. If a certain job costs \$105.50 and lasts more than 1 hour, which of the following expressions represents the length of the job in minutes?

(A) $\dfrac{105.50 - y + z}{z}$

(B) $\dfrac{105.50 - y + 60z}{y}$

(C) $\dfrac{105.50 - y}{z}$

(D) $\dfrac{105.50 - y - z}{z} + 60$

(E) $\dfrac{105.50 - y}{z} + 60$

Answers and Explanations

Answer Key:

2. (C)	**7.** (D)	**15.** (A)
6. (B)	**12.** (C)	**18.** (E)
6. (C)	**13.** (B)	**20.** (E)

2. If x is an even integer, and y is an odd integer, which of the following is an even integer?

(A) $3x + y$
(B) $(x + y)^2$
(C) $2y + x^2$
(D) $2x + 3y$
(E) $3(x + y)$

Explanation:
Plug in an even integer for x and an odd integer for y. There is nothing to box here. We just work through the answer choices to find an even integer.
Let $x = 2$ and $y = 3$

(A) $3x + y$
  $3(2) + 3 =$
    $6 + 3 = 9$ (not even)

(D) $2x + 3y$
  $2(2) + 3(3) =$
    $4 + 9 = 13$ (not even)

(B) $(x + y)^2$
  $(2 + 3)^2 =$
    $5^2 = 25$ (not even)

(E) $3(x + y)$
  $3(2 + 3) =$
    $3(5) = 15$ (not even)

(C) $2y + x^2$
  $2(3) + 2^2 =$
    $6 + 4 = 10$ ✓ (even)

---

6. If the average of a and $a + 4$ is b, and if the median of $a, a + 2, b$, and $b - 2$ is c, what is the average of a and b, in terms of c?

(A) 2
(B) c
(C) $\frac{c}{2}$
(D) $c + \frac{3}{2}$
(E) $c + 2$

Explanation:
Let $a = 2$
If $a = 2$, then $a + 4 = 2 + 4 = 6$
The average of a and $a + 4$ is b. So, b is the average of 2 and 6:

$$\text{Average} = \frac{\text{Total}}{\text{\# of things}}$$

$$b = \frac{2 + 6}{2}$$

$$b = 4$$

The median of $a, a + 2, b$, and $b - 2$ is c
So, c is the median of the following set of numbers: $2, 2 + 2, 4, 4 - 2$, or $2, 4, 4, 2$
Rearranged: $2, 2, 4, 4$

To find the median, find the average of the two middle numbers: $\frac{2 + 4}{2} = 3$

The median is 3, so $c = 3$

The average of a and b is $\frac{2+4}{2} = 3$

$\boxed{3}$

Plug in 3 for c to each of the answer choices to find your boxed answer:

(A) 2   (D) $c + \frac{3}{2}$

(B) c

    3                $3 + \frac{3}{2} = \frac{6}{2} + \frac{3}{2} = \frac{9}{2} = 4\frac{1}{2}$

(C) $\frac{c}{2}$ (E) $c + 2$

                         $3 + 2 = 5$

    $\frac{3}{2}$

---

6. Bob is three times as old as Jenny, who is five times as old as Lou. If Bob is x years old, then in ten years how old will Lou be in terms of x?

(A) $\dfrac{x}{5+10}$

(B) $\dfrac{x+2}{2}$

(C) $\dfrac{x+150}{15}$

(D) $\dfrac{3x}{5+10}$

(E) $\dfrac{3x}{3+10}$

Explanation:
Plug in for Jenny's age first, because it gets the most action.
Jenny's age $(J) = 10$

Bob is three times as old as Jenny, so Bob's age $(B) = 3 \times 10 = 30$
Bob's age is x, so $x = 30$

Jenny's age is 5 times Lou's age, so Lou's age $(L) = 10 \div 5 = 2$
$L = 2$

In ten years, how old will Lou be:
$10 + L =$
$10 + 2 = 12$ $\boxed{12}$

Plug in 30 to the answer choices to find a match for your boxed value:

(A) $\dfrac{x}{5+10} = \dfrac{30}{5+10} = \dfrac{30}{15} = 2$

(B) $\dfrac{x+2}{2} = \dfrac{30+2}{2} = \dfrac{32}{2} = 16$

(C) $\dfrac{x+150}{15} = \dfrac{30+150}{15} = \dfrac{180}{15} = 12 \leftarrow$

(D) $\dfrac{3x}{5+10} = \dfrac{3 \cdot 30}{15} = \dfrac{90}{15} = 6$

(E) $\dfrac{3x}{3+10} = \dfrac{3 \cdot 30}{13} = \dfrac{90}{13} = 6\frac{12}{13}$

---

7. Each of the following is equivalent to

$$\frac{x}{xy(xy^2 + z)} \quad EXCEPT$$

(A) $\frac{1}{y(z + xy^2)}$

(B) $\frac{x}{x^2y^3 + xyz}$

(C) $\frac{1}{xy^3 + yz}$

(D) $\frac{1}{y^3 + yz}$

(E) $\frac{1}{y(xy^2 + z)}$

Explanation:
Plug in numbers for $x, y,$ and z and work it out arithmetically!
Let $x = 4, y = 2, z = 1$

$$\frac{x}{xy(xy^2 + z)}$$

$$\frac{4}{4 \cdot 2(4(2)^2 + 1)} = \frac{4}{8(4(4) + 1)} = \frac{4}{8(17)} = \frac{4}{136} = \frac{1}{34} \quad \boxed{\frac{1}{34}}$$

Plug your values for x, y and z into the answer choices to find the one that *doesn't* match our boxed answer:

(A) $\frac{1}{y(z + xy^2)} = \frac{1}{2(1 + 4(2)^2)} = \frac{1}{2(1 + 16)} = \frac{1}{2(17)} = \frac{1}{34}$

(B) $\frac{x}{x^2y^3 + xyz} = \frac{4}{(4)^2(2)^3 + (4)(2)(1)} = \frac{4}{(16)(8) + 8} = \frac{4}{128 + 8} = \frac{4}{136} = \frac{1}{34}$

(C) $\frac{1}{xy^3 + yz} = \frac{1}{4(2)^3 + 2(1)} = \frac{1}{4(8) + 2} = \frac{1}{34}$

(D) $\frac{1}{y^3 + yz} = \frac{1}{(2)^3 + 2(1)} = \frac{1}{8 + 2} = \frac{1}{10} \neq \frac{1}{34}$

(E) $\frac{1}{y(xy^2 + z)} = \frac{1}{2(4(2)^2 + 1)} = \frac{1}{2(4(4) + 1)} = \frac{1}{2(17)} = \frac{1}{34}$

---

12. If y is a positive integer and $3^y + 3^{(y+1)} = z$, what is $3^{(y+1)}$ in terms of z?

(A) z^2

(B) $3z + 1$

(C) $\frac{3z}{4}$

(D) $\frac{(3-z)}{2}$

(E) $\frac{2z}{3}$

Explanation:
Plug in $y = 2$

$$3^y + 3^{(y+1)} = z$$
$$3^2 + 3^{(2+1)} = z$$
$$9 + 3^3 = z$$
$$9 + 27 = z$$
$$36 = z$$

So we've found z but the question asks for the value of $3^{(y+1)}$. Plug in 2 for y once again:

$3^{(y+1)}$
$3^{(2+1)}$
$3^3 = 27$ $\quad \boxed{27}$

Plug in your z to the answer choices to find the one that matches your boxed value:

$\boxed{27}$

(A) z^2

$36^2 = 1296$

(B) $3z + 1$

$3(36) + 1 = 109$

(C) $\frac{3z}{4}$

$\frac{3(36)}{4} = \frac{108}{4} = 27$

(D) $\frac{3-z}{2} =$

$\frac{3-36}{2} = \frac{-33}{2} = -16\frac{1}{2}$

(E) $\frac{2z}{3} =$

$\frac{2(36)}{3} = \frac{72}{3} = 24$

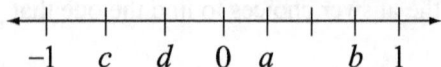

$$-1 \quad c \quad d \quad 0 \quad a \quad b \quad 1$$

13. The letters $c, d, a,$ and b represent numbers on the number line as shown above. Which of the following expressions has the least value?

(A) $c + b$

(B) $c + a$

(C) $d + a$

(D) $c - d$

(E) $b - a$

Explanation:

Plug in values for the variables, using the number line as a guide, to find the least value.

Since there are 2 tick marks between 0 and -1 we can break that half of the number line into thirds. $-1 = \frac{-3}{3},\ c = -\frac{2}{3},\ $ and $\ d = -\frac{1}{3}$

Since there are 3 tick marks between 0 and 1 we can break the right side of the number line into fourths. $1 = \frac{4}{4},\ b = \frac{3}{4},\ a = \frac{1}{4}$

Take these numbers into the answer choices to find the least value.

(A) $c + b$

$-\frac{2}{3} + \frac{3}{4} = -\frac{8}{12} + \frac{9}{12} = \frac{1}{12}$

(B) $c + a$

$-\frac{2}{3} + \frac{1}{4} = -\frac{8}{12} + \frac{3}{12} = -\frac{5}{12}$

(C) $d + a$

$-\frac{1}{3} + \frac{1}{4} = -\frac{4}{12} + \frac{3}{12} = -\frac{1}{12}$

(D) $c - d$

$-\frac{2}{3} - \left(-\frac{1}{3}\right) = -\frac{2}{3} + \frac{1}{3} = -\frac{1}{3}$

(E) $b - a$

$\frac{3}{4} - \frac{1}{4} = \frac{2}{4} = \frac{1}{2}$

$-\frac{5}{12}$ is the least value, so the answer is (B) $c + a$.

---

15. If $2n + 3m = m$, then in terms of n, which of the following must equal $4n + 4m$?

(A) 0

(B) 8

(C) $4n$

(D) $8n$

(E) $6n - 6$

38

Explanation:

Plug in a value for n and solve for m. Let's say $n = 3$:

$$2n + 3m = m$$
$$2(3) + 3m = m$$
$$6 + 3m = m$$
$$6 + 3m - 3m = m - 3m$$
$$6 = -2m$$
$$\frac{6}{-2} = \frac{-2m}{-2}$$
$$-3 = m$$

So we know $m = -3$, but the question asks for which of the answer choices is equal to $4n + 4m$. So, let's plug in our n and m:

$$4n + 4m$$
$$4(3) + 4(-3) =$$
$$12 + (-12) =$$
$$12 - 12 = 0 \quad \boxed{0}$$

Plug in 3 for n in the answer choices:

(A) 0

(B) 8

(C) $4n$
$4(3) = 12$

(D) $8n$
$8(3) = 24$

(E) $6n - 6$
$6(3) - 6$
$18 - 6 = 12$

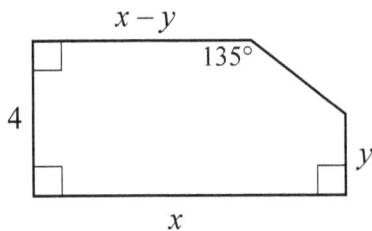

$x - y$

$135°$

4

y

x

18. Which of the following represents the area of the figure above?

(A) $4x + y\sqrt{3}$

(B) $\dfrac{7x}{2} + y\sqrt{3}$

(C) $6x^2 + y\sqrt{3}$

(D) $\dfrac{7x}{2} + y^2$

(E) $4x - 2y + \dfrac{y^2}{2}$

Explanation:

Plug in for x and y. Let $x = 8$ and $y = 2$.

Notice that the shape is a rectangle with a triangular shaped piece missing from its corner.

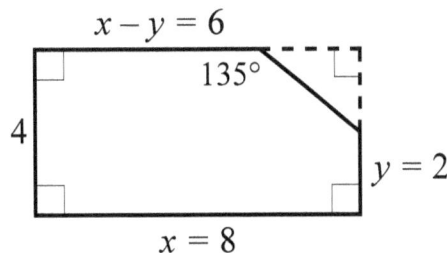

$x - y = 6$

$135°$

4

$y = 2$

$x = 8$

Draw in the missing corner of the rectangle, which is actually a 45°–45°–90° triangle. We know this because 135° and the angle of the triangle sit on a straight line. There are 180° in a straight line.

$180° - 135° = 45°$.

There are 180° in the triangle, so the other angle must also be 45°.

$90° + 45° + 45° = 180°$

We know that the length of the rectangle is 8, so $8 - 6 = 2$. The leg of our triangle is 2. The legs of a 45°–45°–90° triangle are always equal, so we now have a base and a height of 2.

Area $= \frac{1}{2}bh$

$\quad = \frac{1}{2} \cdot 2 \cdot 2 = 2$

The area of the rectangle is length times width; $8 \times 4 = 32$.

The only remaining step is to subtract the area of our triangle from the area of our rectangle.

$32 - 2 = 30$ $\boxed{30}$

Plug our x and y values into the answer choices:

(A) $4x + y\sqrt{3}$
$4(8) + 2\sqrt{3} =$
$32 + 2\sqrt{3} \neq 30$

(B) $\frac{7x}{2} + y\sqrt{3}$
$\frac{7(8)}{2} + 2\sqrt{3} =$
$\frac{56}{2} + 2\sqrt{3} =$
$28 + 2\sqrt{3} \neq 30$

(C) $6x^2 + y\sqrt{3}$
$6(8)^2 + 2\sqrt{3}$
$6(64) + 2\sqrt{3} =$
$384 + 2\sqrt{3} \neq 30$

(D) $\frac{7x}{2} + y^2$
$\frac{7(8)}{2} + 2^2 =$
$\frac{56}{2} + 4 = 28 + 4 =$
$28 + 4 =$
$32 \neq 30$

(E) $4x - 2y + \frac{y^2}{2}$
$4(8) - 2(2) + \frac{2^2}{2}$
$32 - 4 + \frac{4}{2}$
$32 - 4 + 2$
$28 + 2$
$30 \longleftarrow$

20. A plumber charges y dollars for the first hour of work and charges for any additional time at the rate of z dollars per minute. If a certain job costs \$105.50 and lasts more than 1 hour, which of the following expressions represents the length of the job in minutes?

(A) $\dfrac{105.50 - y + z}{z}$

(B) $\dfrac{105.50 - y + 60z}{y}$

(C) $\dfrac{105.50 - y}{z}$

(D) $\dfrac{105.50 - y - z}{z} + 60$

(E) $\dfrac{105.50 - y}{z} + 60$

Explanation:
Plug in for y and z; let $y = 100$, and $z = 2$

Total cost of the job = 105.z0
Cost for the first hour = $y = 100$

Cost of the additional minutes = Total cost – cost of first hour
$$= 105.50 - 100$$
$$= 5.50$$

Cost of each additional minute = $z = 2$

Number of additional minutes = $\dfrac{\text{total cost of additional minutes}}{\text{cost of each additional minute}}$

$$= \dfrac{5.50}{2} = 2.75$$

Length of the job in minutes = 1st hour (60 minutes) + additional minutes
$$= 60 + 2.75$$
$$= 62.75 \quad \boxed{62.75}$$

Plug in the answer choices to find the one that matches your boxed value:

(A) $\dfrac{105.50 - y + z}{z}$

$\dfrac{105.50 - 100 + 2}{2}$

$\dfrac{7.50}{2} = 3.75$

(B) $\dfrac{105.50 - y + 60z}{y}$

$\dfrac{105.50 - 100 + (60 \times 2)}{100}$

$\dfrac{125.50}{100} = 1.255$

(C) $\dfrac{105.50 - y}{z}$

$\dfrac{105.50 - 100}{2}$

$\dfrac{5.50}{2} = 2.75$

(D) $\dfrac{105.50 - y - z}{z} + 60$

$\dfrac{105.50 - 100 - 2}{2} + 60$

$\dfrac{3.50}{2} + 60$

$1.75 + 60 = 61.75$

(E) $\dfrac{105.50 - y}{z} + 60$

$\dfrac{105.50 - 100}{2} + 60$

$\dfrac{5.50}{2} + 60$

$2.75 + 60 = 62.75$

Chapter 3
Must Be/Could Be Problems

Now that we've mastered the Basic Plug Ins, let's look at some Must Be/Could Be problems. The Must Be/Could Be problem is a slight variation of the Basic Plug In. Your tip off to these types of problems is easy: they contain the expression **"which of the following must be"** or **"which of the following could be."** Look for Roman numeral problems in particular and, of course, variables in the answer choices.

Easy numbers like 2, 4, and 5 are still good choices for Must Be/Could Be problems. Unlike on the Basic Plug Ins, however, on Must Be/Could Be problems **0 and 1 are great numbers to plug in**, as are **negatives and fractions**. Think outside the box!

Let's go through some together:

14. If $k - p > k$, which of the following must be true?

(A) $k > p$

(B) $p > k$

(C) $k = p$

(D) $k < 0$

(E) $p < 0$

Explanation:

Notice those variables in the answer choices and the "phrase must be true."

Plug in $k = -3$ and p = -1.

Check the restriction: $k - p > k$

$$-3 - (-1) > -3$$
$$-3 + 1 > -3$$
$$-2 > -3 \checkmark$$

Since the restriction works, go through the answer choices, plugging in our values for k and p:

(A) $k > p$ (D) $k < 0$
    $-3 > -1$ $-3 < 0$

(B) $p > k$ (E) $p < 0$
    $-1 > -3$ $-1 < 0$

(C) $k = p$
    $-3 = -1$

So we can eliminate (A) and (C). They weren't true based on the numbers we plugged in, which means they must NOT ALWAYS be true.

Now choose new values to plug in, say, $k = 3$ and $p = -3$.

Check the restriction: $k - p > k$

$$3 - (-3) > 3$$
$$3 + 3 > 3$$
$$6 > 3 \checkmark$$

Since the restriction works, go through the remaining answer choices:

(B) $p > k$
    $-3 > 3$

(D) $k < 0$
    $3 < 0$

(E) $p < 0$
    $-3 < 0$ $\checkmark$

Eliminate (B) and (D), and the correct answer is (E).

Answer: (E) $p < 0$

Here's another one; this time a Roman numeral problem:

16. If the expression $2x^2 + 3y$ is odd, and x and y are integers, which of the following statements must be true?

   I. y is odd
  II. x is odd
 III. $y - x$ is odd

(A) I only
(B) III only
(C) I and III
(D) II and III
(E) I, II, and III

Explanation:
Plug in $x = 2$ and plug in the odd number 5 as the sum of $2x^2 + 3y$. Solve for y:

$$2x^2 + 3y = 5$$
$$2(2)^2 + 3y = 5$$
$$2(4) + 3y = 5$$
$$8 + 3y = 5$$
$$8 + 3y - 8 = 5 - 8$$
$$3y = -3$$
$$\frac{3y}{3} = \frac{-3}{3}$$
$$y = -1$$

So, $y = -1$. Now, go through the Roman numerals and eliminate the ones we can:

  I. y is odd
    -1 is odd
 ~~II.~~ x is odd
    2 is NOT odd
 III. $y - x$ is odd
    $-1 - 2$ is odd
    -3 is odd

Eliminate (D) and (E), because they contain II.

This leaves us with I and III working. But don't stop there – we don't know that they MUST BE true. We plugged in an even number, 2, for x, so let's plug in an odd number for x, like 3. Let's set $2x^2 + 3y$ equal to 9 and solve for y.

$$2x^2 + 3y = 9$$
$$2(3)^2 + 3y = 9$$
$$2(9) + 3y = 9$$
$$18 + 3y = 9$$
$$18 + 3y - 18 = 9 - 18$$
$$3y = -9$$
$$\frac{3y}{3} = \frac{-9}{3}$$
$$y = -3$$

  I. y is odd
    -3 is odd
 ~~III.~~ $y - x$ is odd
    $-3 - 3$ is odd
    -6 is NOT odd

Only I remains.

Answer: (A) I only

Let's see how a Must Be problem works on Geometry:

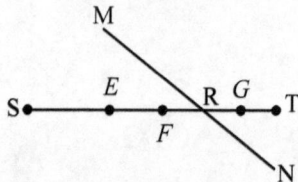

8. In the figure, $\overline{MN}$ intersects $\overline{ST}$ at point R. If E, F, and G are midpoints of $\overline{SR}$, $\overline{ST}$, and $\overline{RT}$, respectively, which of the following must be true?

 I. The distance from S to F is greater than the distance from R to E.
 II. The distance from F to G is greater than the distance from R to E.
 III. The distance from S to E is greater than the distance from T to G.

(A) I only
(B) II only
(C) I and III only
(D) II and III only
(E) I, II, and III

Explanation:
Plug in values for the distances. Let's say, $\overline{ST} = 8$, and $\overline{SR} = 6$. That leaves $\overline{RT} = 2$.

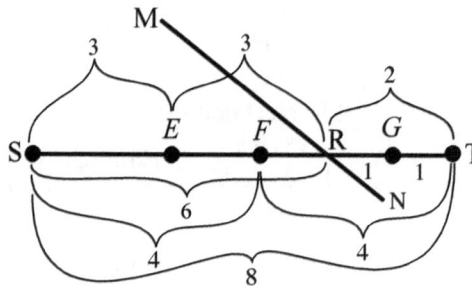

Since E is the midpoint of $\overline{SR}$, $\overline{SE}$ and $\overline{ER}$ both equal 3. Since F is the midpoint of $\overline{ST}$, $\overline{SF}$ and $\overline{FT}$ both equal 4. Since G is the midpoint of $\overline{RT}$, $\overline{RG}$ and $\overline{GT}$ both equal 1.

Plug in your values to the Roman numerals:

I. The distance from S to F is greater than the distance from R to E.
$4 > 3$ ✓

II. The distance from F to G is greater than the distance from R to E.
$(4 - 1) > 3$

$3 > 3$ ✗
Eliminate (B), (D) and (E), because they contain II.

III. The distance from S to E is greater than the distance from T to G.
$3 > 1$ ✓
Eliminate (A), because it doesn't contain III. Both I and III work!

Answer: (C) I and III only

44

The Could Be problems are a little bit different.

17. If t is one of 3 consecutive positive even integers whose sum is v, which of the following could be true?

(A) $3t - 6 = v$

(B) $3t + 3 = v$

(C) $3t - 3 = v$

(D) $6n + 1 = v$

(E) $7t = v$

Explanation:

Plug in the consecutive integers 2, 4 and 6. Their sum is 12, so $v = 12$. Let $t = 2$.

Step 1:

Plug in 2 to the answers and we can see that none of them work:

(A) $3t - 6 = v$

$3(2) - 6 =$

$6 - 6 = 0$

$0 \neq 12$

(B) $3t + 3 = v$

$3(2) + 3 =$

$6 + 3 = 9$

$9 \neq 12$

(C) $3t - 3 = v$

$3(2) - 3 =$

$6 - 3 = 3$

$3 \neq 12$

(D) $6n + 1 = v$

$6(2) + 1 =$

$12 + 1 = 13$

$13 \neq 12$

(E) $7t = v$

$7(2) = 14$

$14 \neq 12$

Step 2:

But it's a Could Be question, which means that the right answer doesn't have to work all the time; all you need to do is get it to work once!

Try $t = 4$

(A) $3t - 6 = v$

$3(4) - 6 =$

$12 - 6 = 6$

$6 \neq 12$

(B) $3t + 3 = v$

$3(4) + 3 =$

$12 + 3 = 15$

$15 \neq 12$

(C) $3t - 3 = v$

$3(4) - 3 =$

$12 - 3 = 9$

$9 \neq 12$

(D) $6n + 1 = v$

$6(4) + 1 =$

$24 + 1 = 25$

$25 \neq 12$

(E) $7t = v$

$7(4) = 28$

$28 \neq 12$

Step 3:

4 doesn't work either, so now let $t = 6$:

(A) $3t - 6 = v$

$3(6) - 6 =$

$18 - 6 = 12$

$12 = 12$ ✓

(B) $3t + 3 = v$

$3(6) + 3 =$

$18 + 3 = 21$

$21 \neq 12$

(C) $3t - 3 = v$

$3(6) - 3 =$

$18 - 3 = 15$

$15 \neq 12$

(D) $6n + 1 = v$

$6(6) + 1 =$

$36 + 1 = 37$

$37 \neq 12$

(E) $7t = v$

$7(6) = 42$

$42 \neq 12$

Answer: (A) $3t - 6 = v$

Let's see how you do on your own with this drill of 5 questions.

Must Be/Could Be Drill

Some of Mark's sisters sing.
None of Billy's sisters dance.

9. If the two statements above are true, which of the following statements must also be true?

  I. Billy's sisters sometimes sing.
  II. Mark and Billy's sisters never dance together.
  III. Mark's sisters never dance.

  (A) I only
  (B) II only
  (C) III only
  (D) I and II only
  (E) II and III only

10. In $\triangle ABC$, $\overline{AB} = 8$ and $\overline{BC} = 6$. Which of the following could be the value of $\overline{AC}$?

  I. 10
  II. 12
  III. 14

  (A) I only
  (B) II only
  (C) III only
  (D) I and II only
  (E) II and III only

13. If n is an integer, which of the following must be true about $2n + 5$?

  (A) It is divisible by 7.
  (B) It is greater than n.
  (C) It is greater than 5.
  (D) It is odd.
  (E) It is even.

16. For all numbers p and q, let $p \lozenge q$ be defined by $p \lozenge q = p - q + pq$. For all numbers a and b, which of the following must be true?

  I. $a \lozenge b = b \lozenge a$
  II. $(b - 2) \lozenge b = (a \lozenge a) - 2$
  III. $(a \lozenge b) - (b \lozenge a) = (2 \lozenge a) - (2 \lozenge 2b - a)$

  (A) I only
  (B) II only
  (C) III only
  (D) I and III only
  (E) II and III only

18. On a number line, the distance between point d and point f is greater than 50. Which of the following expressions must be true?

  A) $d - f > 50$
  B) $d + f > 50$
  C) $|d - f| > 50$
  D) $|d + f| > 50$
  E) $|d| \bullet |f| > 50$

Answers and Explanations

Some of Mark's sisters sing.
None of Billy's sisters dance.

9. If the two statements above are true, which of the following statements must also be true?

  I. Billy's sisters sometimes sing.
 II. Mark and Billy's sisters never dance together.
III. Mark's sisters never dance.

(A) I only
(B) II only
(C) III only
(D) I and II only
(E) II and III only

Explanation:
  I. – All we know about Billy's sisters is that they don't dance; we have no idea if any of them sing. They may, but it is not true beyond a doubt.
 II. – None of Billy's sisters dance, so Mark and Billy's sisters will not be dancing together if Billy's sisters do not dance. So II must be true.
III. – All we know is that some of Mark's sisters sing, but we have no way of knowing if they dance or not. So, III need not be true.

---

10. In $\triangle ABC$, $\overline{AB} = 8$ and $\overline{BC} = 6$. Which of the following could be the value of $\overline{AC}$?

  I. 10
 II. 12
III. 14

(A) I only
(B) II only
(C) III only
(D) I and II only
(E) II and III only

Explanation:
We do not know the degree measures of the angles, which means all we have to work with is the **Third Side Triangle Rule**, which states that the sum of any two sides of a triangle must be greater than the third side.

Add the two given sides: $8 + 6 = 14$

Then subtract: $8 - 6 = 2$

So: $2 < \overline{AC} < 14$

  I. $2 < 10 < 14$ ✓
 II. $2 < 12 < 14$ ✓
III. $2 < 14 \not< 14$ ✗

---

13. If n is an integer, which of the following must be true about $2n + 5$?

(A) It is divisible by 7.
(B) It is greater than n.
(C) It is greater than 5.
(D) It is odd.
(E) It is even.

Explanation:
First, try plugging in $n = 2$
So, $2n + 5 =$
$$2(2) + 5 = 9$$

(A) 9 is divisible by 7
(B) 9 is greater than 2
(C) 9 is greater than 5
(D) 9 is odd
(E) 9 is even

We can eliminate (A) and (E).

Now, since we've tried a positive even integer, let's be creative and plug in a negative odd integer, such as –5.
So, $2n + 5 =$
$$2(-5) + 5 = -5$$

(B) –5 is greater than –5
(C) –5 is greater than 5
(D) –5 is odd

Eliminate (B) and (C) and we are left with (D).

---

16. For all numbers p and q, let $p \lozenge q$ be defined by $p \lozenge q = p - q + pq$. For all numbers a and b, which of the following must be true?

I. $a \lozenge b = b \lozenge a$
II. $(b - 2) \lozenge b = (a \lozenge a) - 2$
III. $(a \lozenge b) - (b \lozenge a) = (2 \lozenge a) - (2 \lozenge 2b - a)$

(A) I only
(B) II only
(C) III only
(D) I and III only
(E) II and III only

Explanation:
Plug in $a = 2$ and $b = 3$. Plug these numbers into the answers according to the function definition given. Remember, a is in the p position and b is in the q position.

I. $a \lozenge b = b \lozenge a$
$2 \lozenge 3 = 3 \lozenge 2$
$2 - 3 + 2 \cdot 3 = 3 - 2 + 3 \cdot 2$
$-1 + 6 = 1 + 6$
$5 \neq 7$
Eliminate (A) and (D), because they contain I.

II. $(b - 2) \lozenge b = (a \lozenge a) - 2$
$(3 - 2) \lozenge 3 = (2 \lozenge 2) - 2$
$1 \lozenge 3 = (2 \lozenge 2) - 2$
$1 - 3 + 1(3) = [2 - 2 + 2(2)] - 2$
$-2 + 3 = [0 + 4] - 2$
$1 = [4] - 2$
$1 \neq 2$
Eliminate (B) and (E), because they contain II.

Notice how only (C) is left. No need to work out the math for III, but here is how it works in case you are curious.

III. $(a \Diamond b) - (b \Diamond a) = (2 \Diamond a) - (2 \Diamond 2b - a)$

$(2 \Diamond 3) - (3 \Diamond 2) = (2 \Diamond 2) - (2 \Diamond 2(3) - 2)$

$(2 \Diamond 3) - (3 \Diamond 2) = (2 \Diamond 2) - (2 \Diamond 4)$

$2 - 3 + 6 - [3 - 2 + 6] = 2 - 2 + 4 - [2 - 4 + 8]$

$-1 + 6 - [1 + 6] = 0 + 4 - [-2 + 8]$

$5 - [7] = 4 - [6]$

$-2 = -2$

---

18. On a number line, the distance between point d and point f is greater than 50. Which of the following expressions must be true?

(A) $d - f > 50$
(B) $d + f > 50$
(C) $|d - f| > 50$
(D) $|d + f| > 50$
(E) $|d| \bullet |f| > 50$

Explanation:
Plug in $d = 0$ and $f = 51$. Go through the answer choices:

(A) $d - f > 50$
$0 - 51 > 50$
$-51 > 51$

(B) $d + f > 50$
$0 + 51 > 50$
$51 > 50$

(C) $|d - f| > 50$
$|0 - 51| > 50$
$|-51| > 50$
$51 > 50$

(D) $|d + f| > 50$
$|0 + 51| > 50$
$|51| > 50$
$51 > 50$

(E) $|d| \bullet |f| > 50$
$|0| \bullet |51| > 50$
$0 \bullet 51 > 50$
$0 > 50$

We can eliminate (A) and (E). It doesn't specify that d and f have to be positive integers, so plug in $f = -50$ and $d = 1$.

(B) $d + f > 50$
$1 + (-50) > 50$
$-49 > 50$

(D) $|d + f| > 50$
$|1 + (-50)| > 50$
$|-49| > 50$
$49 > 50$

(C) $|d - f| > 50$
$|1 - (-50)| > 50$
$|51| > 50$
$51 > 50$ ✓

Now we can eliminate (B) and (D) and we're left with Answer (C).

Chapter 4
Answer Choice Test (A.C.T.)

The next type of Plugging In occurs when there are **numbers in the answer choices** and **the question asks for a single, specific thing** like "What is *x*?" "How old is Bob?" or " How many marbles does Lou have?" One of the answer choices has to be the number of marbles Lou has; why not take the answer choices back into the question and turn it into a simple arithmetic problem? So when you have the urge to set up an algebraic equation, STOP and remember to **A.C.T.** by plugging in the answers - to the **A**nswer **C**hoice **T**est! Let's see how it works on the following basic problem.

9. A movie theatre made up of 23 rows can seat
a total of 221 people. If some of the rows can
seat 9 people and the others can seat 10 people,
how many rows seat 9 people?

(A) 6
(B) 7
(C) 8
(D) 9
(E) 10

Explanation:
Notice how there are numbers in the answer choices and the question asks for a single specific thing: *how many rows seat 9 people?* As you can see, the answer choices are arranged in order from least to greatest. The answer choices will ALWAYS be in either ascending or descending order so we plug in (C) first. Throwing in (C) ultimately saves time, because even if it is not the correct answer, you will often be able to tell if you want a bigger or a smaller number.

So plug in (C) 8:
If 8 rows seat 9 people and there are 23 rows total,
23 – 8 = 15, so there are 15 rows that seat 10 people
8 rows × 9 people = 72 people and 15 rows × 10 people = 150 people
72 + 150 = 222
Since there is supposed to be a total of 221, we need a larger number of rows with only 9 people (and so fewer rows with 10).

Let's try (D) 9:
If 9 rows seat 9 people and there are 23 rows total,
23 – 9 = 14, so there are 14 rows that seat 10 people
9 rows × 9 people = 81 and 10 rows × 14 people = 140 people
140 + 81 = 221

By using the Answer Choice Test, we only had to try two of the choices before arriving at the correct answer (D) 9.

Answer: (D) 9

───────────────

Let's try another one:

50

8. If $(y - 4)^2 = 49$ and $y > 0$, what is the value of y?

(A) 2
(B) 3
(C) 5
(D) 7
(E) 11

Explanation:
Notice how there are numbers in the answer choices and the question asks for a single specific thing: *what is the value of y*? Remember to plug in (C) first:

(C) 5
$(y - 4)^2 = 49$
$(5 - 4)^2 = 49$
$1^2 = 49$
$1 \neq 49$

We need a much bigger number so let's jump to (E) 11

(E) 11
$(y - 4)^2 = 49$
$(11 - 4)^2 = 49$
$7^2 = 49$
$49 = 49$ ✓

Answer: (E) 11

---

8. Three consecutive even integers are such that
four times the smallest is two times the largest.
What is the largest of these integers?

(A) –2
(B) 0
(C) 4
(D) 8
(E) 10

Explanation:
Notice how there are numbers in the answer choices and the question asks for a single specific thing. However, this problem demonstrates an exception to the "always start with (C) rule" when using the Answer Choice Test. Whenever ETS asks for the smallest or largest start with the smallest or largest answer choice. The reason for this is that (C) might just work, but might not be the smallest or largest possibility in the answer choices. So let's start with (E):

(E) 10
The question asks for the largest of three consecutive even integers, so the sequence would be: 6, 8, 10
Four times the smallest: $6 \times 4 = 24$
Two times the largest: $2 \times 10 = 20$
$$20 \neq 24$$

Move to (D) 8
The sequence would be: 4, 6, 8
Four times the smallest: $4 \times 4 = 16$
Two times the largest: $2 \times 8 = 16$
$$16 = 16 ✓$$

Answer: (D) 8

We can, of course, plug in the answer choices on Geometry questions too. Let's see how this works.

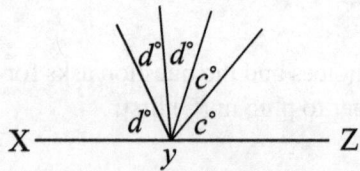

Note: Figure not drawn to scale

4. In the figure, point y lies on $\overline{XZ}$. If d and c are both integers, what is one possible value of c?

(A) 32
(B) 36
(C) 40
(D) 50
(E) 55

Explanation:
Do the A.C.T.
Start by plugging in (C) 40
$c = 40$
There are 3 ds and 2 cs making up the straight line (and there are 180° in a straight line), so:
$3d + 2c = 180°$
$3d + 2(40) = 180$
$3d + 80 = 180$
$3d + 80 - 80 = 180 - 80$
$3d = 100$
$\dfrac{3d}{3} = \dfrac{100}{3}$
$d = 33.3$
33.3 is not an integer.

Now try (B) 36
$c = 36$
$3d + 2c = 180°$
$3d + 2(36) = 180$
$3d + 72 = 180$
$3d + 72 - 72 = 180 - 72$
$3d = 108$
$\dfrac{3d}{3} = \dfrac{108}{3}$
$d = 36$
36 is an integer. ✓

Answer: (B) 36

Let's do a number 20 problem together, which is supposedly one of the hardest problems on a math section.

$$k(n) = (n - 1)^2 - 30n + c$$

20. The function k above represents the number of people at the Sherwood Mall on any given day in 2009. If n represents the day of the year and c is a constant, on what number day was the number of people at the mall the same as it was on day number 11?

(A) 21
(B) 31
(C) 41
(D) 51
(E) 61

Explanation:
The answer choices stand for the day number. First let's find out how many people were at the mall on day number 11.
Plug in 11 for n:

$k(n) = (n - 1)^2 - 30n + c$

$k(11) = (11 - 1)^2 - (30 \times 11) + c$

$k(11) = (10)^2 - 330 + c$

$k(11) = 100 - 330 + c$

$k(11) = -230 + c$

$$\boxed{-230 + c}$$

Don't worry about the c. It's a constant, which means it's the same number no matter the day and so isn't affecting the problem.

Plug in (C) 41
$k(n) = (n - 1)^2 - 30n + c$
$k(41) = (41 - 1)^2 - (30 \times 41) + c$
$k(41) = (40)^2 - 1230 + c$
$k(41) = 1600 - 1230 + c$
$k(41) = 370 + c$
Cross out (C) because $370 + c \neq -230 + c$

Plug in (B) 31
$k(n) = (n - 1)^2 - 30n + c$
$k(31) = (31 - 1)^2 - (30 \times 31) + c$
$k(31) = (30)^2 - 930 + c$
$k(31) = 900 - 930 + c$
$k(31) = -30 + c$
Cross out (B) because $-30 + c \neq -230 + c$

Plug in (A) 21
$k(n) = (n - 1)^2 - 30n + c$
$k(21) = (21 - 1)^2 - (30 \times 21) + c$
$k(21) = (20)^2 - 630 + c$
$k(21) = 400 - 630 + c$
$k(21) = -230 + c$

Answer: (A) 21

Time for an A.C.T. drill.

A.C.T. Drill

2. If 36 is the average of y, $4y$, and $7y$, what is the value of y?

(A) 5
(B) 6
(C) 7
(D) 8
(E) 9

4. If 32 out of 128 sophomores at Sunnybrook High School are members of the environmental club, and if the ratio of the number of sophomores who are members of the environmental club is equal to the ratio of the number of freshmen who are members of the environmental club, how many of the 132 freshmen are members of the environmental club?

(A) 33
(B) 30
(C) 28
(D) 25
(E) 20

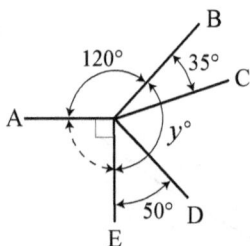

6. In the figure above, what is the value of y?

(A) 65
(B) 100
(C) 115
(D) 150
(E) 165

8. April collected quarters and dimes over the course of one year. When she totaled up the savings in her piggy bank at the end of the year she found she had collected a total of $32.95 worth of quarters and dimes. Which of the following could be the number of dimes that April collected?

(A) 86
(B) 92
(C) 95
(D) 104
(E) 120

12. If $\frac{z^3}{w}$ is a negative integer, but $\frac{z}{w}$ is NOT an integer, which of the following could be the values of z and w?

(A) $z = -1, w = 1$
(B) $z = -7, w = -3$
(C) $z = 4, w = -2$
(D) $z = 6, w = 4$
(E) $z = 6, w = -8$

13. Workers at a car factory work for a total of 9 hours a day, which includes three 16-minute breaks scheduled at equal intervals throughout the workers' shifts. How long, in minutes, is each interval between breaks?

(A) 123
(B) 146
(C) 164
(D) 180
(E) 242

17. Cindy is making a patchwork quilt made up of three different types of squares. One square is the pattern of a horse, one the pattern of a cow, and the other the pattern of a pig. She starts with 3 horses, 4 cows, and 8 pigs and continues the pattern in that order until the quilt is completed. If the last square is that of a horse, what could be the total number of squares that make up the patchwork quilt?

(A) 48
(B) 56
(C) 60
(D) 64
(E) 72

18. If the square of a is equal to 9 times the square of b, and a is 1 more than 3 times b, what is the value of b?

(A) -9

(B) $-\dfrac{1}{6}$

(C) $-\dfrac{1}{3}$

(D) $\dfrac{1}{6}$

(E) $\dfrac{1}{3}$

18. $x, y,$ and z are consecutive positive integers such that $x < y < z$. If the units digit of the sum of x and y is 7, what is the units digit of z?

(A) 4
(B) 3
(C) 2
(D) 1
(E) 0

20. As part of a community gardening project, City A divides a square plot of land into x rows with x squares each. If z of these squares make up the border of the garden, which of the following could be z?

(A) 9
(B) 26
(C) 54
(D) 60
(E) 66

Answers and Explanations

Answer Key:

2. (E)	12. (E)	18. (E)
4. (A)	13. (A)	20. (D)
6. (D)	17. (A)	
8. (B)	18. (B)	

2. If 36 is the average of y, $4y$, and $7y$, what is the value of y?

(A) 5
(B) 6
(C) 7
(D) 8
(E) 9

Explanation:
Start with (C) 7
y, $4y$, and $7y$
7, $4(7)$, and $7(7)$
$7, 28, 49$
Find the average: $\dfrac{7+28+49}{3} = \dfrac{84}{3} = 28$

But $28 \neq 36$

We need a bigger number, so let's jump to (E) 9
y, $4y$, and $7y$
9, $4(9)$, and $7(9)$
$9, 36, 63$
Find the average: $\dfrac{9+36+63}{3} = \dfrac{108}{3} = 36$

$36 = 36$ ✓

---

4. If 32 out of 128 sophomores at Sunnybrook High School are members of the environmental club, and if the ratio of the number of sophomores who are members of the environmental club is equal to the ratio of the number of freshmen who are members of the environmental club, how many of the 132 freshmen are members of the environmental club?

(A) 33
(B) 30
(C) 28
(D) 25
(E) 20

Explanation:
The ratio of sophomores in the club is $\dfrac{32}{128}$, which reduces to $\dfrac{1}{4}$.

Start by plugging in (C) 28:

$\dfrac{28}{132}\left(=\dfrac{7}{33}\right)$ doesn't reduce to $\dfrac{1}{4}$

Next plug in (B) 30:

$\dfrac{30}{132}\left(=\dfrac{5}{22}\right)$ doesn't reduce to $\dfrac{1}{4}$

Now, try (A) 33:

$\dfrac{33}{132}$ reduces to $\dfrac{1}{4}$ ✓

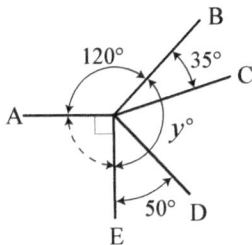

6. In the figure above, what is the value of y?

(A) 65
(B) 100
(C) 115
(D) 150
(E) 165

Explanation:
Plug in (C) 115
$y = 115$
$115 + 120 + 90 = 325$
But the sum of the angles should be 360° - a circle is 360°.
Note: you can disregard the 50° and 35°, because they are already accounted for in $y°$

Try (D) 150
$y = 150$
$150 + 120 + 90 = 360$ ✓

---

8. April collected quarters and dimes over the course of one year. When she totaled up the savings in her piggy bank at the end of the year she found she had collected a total of $32.95 worth of quarters and dimes. Which of the following could be the number of dimes that April collected?

(A) 86
(B) 92
(C) 95
(D) 104
(E) 120

Explanation:
Start with (C) 95 to suppose there are 95 dimes.
Each dime is worth 10 cents, so $95 \times .10 =$ $9.50

And $32.95 - $9.50 = $23.45, leaving us with $23.45, which would have to be made up of quarters, but .25 doesn't divide into the leftover amount evenly.

Now try (B) 92 dimes
$92 \times .10 = $9.20
$32.95 - $9.20 = $23.75, which is evenly divisible by .25 ✓

---

12. If $\frac{z^3}{w}$ is a negative integer, but $\frac{z}{w}$ is NOT an integer, which of the following could be the values of z and w?

(A) $z = -1, w = 1$
(B) $z = -7, w = -3$
(C) $z = 4, w = -2$
(D) $z = 6, w = 4$
(E) $z = 6, w = -8$

Explanation:
Plug in the answers choices, starting with (C):

(C) $z = 4$, $w = -2$

$$\frac{z^3}{w} = \frac{4^3}{-2} = \frac{64}{-2} = -32$$

$$\frac{z}{w} = \frac{4}{-2} = -2 \leftarrow$$

−2 is an integer

(D) $z = 6$, $w = 4$

$$\frac{z^3}{w} = \frac{6^3}{4} = \frac{216}{4} = 54 \nwarrow$$

$$\frac{z}{w} = \frac{6}{4} = \frac{3}{2}$$

54 is not negative

(E) $z = 6$, $w = -8$

$$\frac{z^3}{w} = \frac{6^3}{-8} = \frac{216}{-8} = -27 \ \checkmark \ \text{(negative integer)}$$

$$\frac{z}{w} = \frac{6}{-8} = -\frac{3}{4} \ \checkmark \ \text{(not an integer)}$$

---

13. Workers at a car factory work for a total of 9 hours a day with three 16-minute breaks scheduled at equal intervals throughout the workers' shifts. How long, in minutes, is each interval between breaks?

(A) 123
(B) 146
(C) 164
(D) 180
(E) 242

Explanation:
First note that there are 4 intervals throughout the day:

Interval
Break — 16 minutes
Interval
Break — 16 minutes   } 9 hours
Interval
Break — 16 minutes
Interval

9 hours = 9 × 60 minutes = 540 minutes   540

Plug in (C) 164
Since there are four intervals and three breaks,
we get (164 × 4) + (16 × 3) = 656 + 48 = 704
540 ≠ 704

Since 704 is quite a lot larger than our target, 540,
let's skip to a much smaller number, (A) 123:
(123 × 4) + (16 × 3) = 492 + 48 = 540 ⤺

17. Cindy is making a patchwork quilt made up of three different types of squares. One square is the pattern of a horse, one the pattern of a cow, and the other the pattern of a pig. She starts with 3 horses, 4 cows, and 8 pigs and continues the pattern in that order until the quilt is completed. If the last square is that of a horse, what could be the total number of squares that make up the patchwork quilt?

(A) 48
(B) 56
(C) 60
(D) 64
(E) 72

Explanation:
There are 15 patches in the pattern:

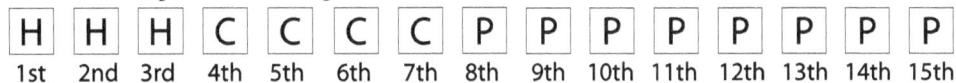

H	H	H	C	C	C	C	P	P	P	P	P	P	P	P
1st	2nd	3rd	4th	5th	6th	7th	8th	9th	10th	11th	12th	13th	14th	15th

Plug in (C) 60 for the total number of squares.
$60 \div 15 = 4$
Since there is no remainder, the last square would be the last in the pattern, which is pig.

Try (D) 64
$64 \div 15 = 4\ R4$
Since the remainder is four, the last square would be the fourth square in the pattern, which is cow.

Try (E) 72
$72 \div 15 = 4\ R12$
Since the remainder is twelve, the last square would be the twelfth square in the pattern, which is pig.

Try (A) 48
$48 \div 15 = 3\ R3$
Since the remainder is three, the last square would be the third square in the pattern, which is horse.

---

18. If the square of a is equal to 9 times the square of b, and a is 1 more than 3 times b, what is the value of b?

(A) -9
(B) $-\frac{1}{6}$
(C) $-\frac{1}{3}$
(D) $\frac{1}{6}$
(E) $\frac{1}{3}$

Explanation:
First translate the problem:

Equation 1:
the square of a is equal to 9 times the square of b

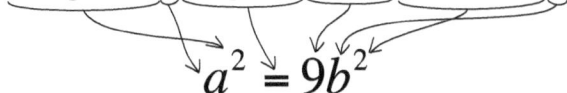

$$a^2 = 9b^2$$

Equation 2:
a is 1 more than 3 times b

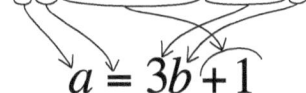

$$a = 3b + 1$$

Plug in (C) $-\frac{1}{3}$

$b = -\frac{1}{3}$

Equation 2:

$a = 3b + 1$

$a = 3\left(-\frac{1}{3}\right) + 1$

$a = -\frac{3}{3} + 1$

$a = -1 + 1$

$a = 0$

Equation 1:

$a^2 = 9b^2$

$0^2 = 9\left(-\frac{1}{3}\right)^2$

$0 = 9\left(-\frac{1}{3} \times -\frac{1}{3}\right)$

$0 = 9\left(\frac{1}{9}\right)$

$0 = \frac{9}{9}$

$0 \neq 1$

Now try (B) $-\frac{1}{6}$

$b = -\frac{1}{6}$

Equation 2:

$a = 3b + 1$

$a = 3\left(-\frac{1}{6}\right) + 1$

$a = -\frac{3}{6} + 1$

$a = -\frac{1}{2} + 1$

$a = \frac{1}{2}$

Equation 1:

$a^2 = 9b^2$

$\left(\frac{1}{2}\right)^2 = 9\left(-\frac{1}{6}\right)^2$

$\left(\frac{1}{2} \times \frac{1}{2}\right) = 9\left(-\frac{1}{6} \times -\frac{1}{6}\right)$

$\frac{1}{4} = 9\left(\frac{1}{36}\right)$

$\frac{1}{4} = \frac{9}{36}$

$\frac{1}{4} = \frac{1}{4}$ ✓

---

18. x, y, and z are consecutive positive integers such that $x < y < z$. If the units digit of the sum of x and y is 7, what is the units digit of z?

(A) 4
(B) 3
(C) 2
(D) 1
(E) 0

Explanation:
Plug in the answer choices, starting with (C) 2
The units digit of z is 2, therefore z could be 12, or 22, or 32, etc. Let's say z is 12.
Since x, y and z are consecutive and $x < y < z$, x would be 10 and y would be 11.
But $10 + 11 = 21$, resulting in a units digit of 1, not 7

(D) 1
The units digit of z is 1, therefore, let's make $z = 11$
So, x, y and z are 9, 10 and 11, respectively.
But $9 + 10 = 19$, resulting in a units digit of 9, not 7

(E) 0
The units digit of z is 0, therefore, let's make $z = 10$
So, x, y and z are 8, 9 and 10, respectively.
And $8 + 9 = 17$, resulting in a units digit of 7

20. As part of a community gardening project, City A divides a square plot of land into x rows and x squares each. If z of these squares make up the border of the garden, which of the following could be z?

(A) 9
(B) 26
(C) 54
(D) 60
(E) 66

Explanation:
Let's first throw in numbers to make sense of the question.
Let's say $x = 4$. Draw the figure, made up of 4 rows of 4 squares:

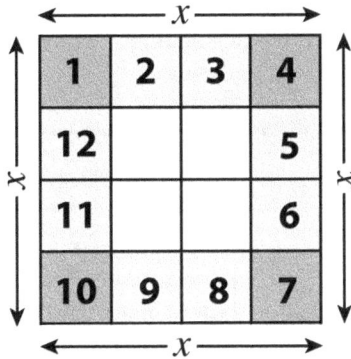

$4x$ gives you all the squares on the perimeter, but you have counted the corners twice! No matter what x is, there will always be four corners. So subtract 4 to get rid of the overlap: $4x - 4 = z$

Now plug in for z, starting with (C) 54:

(C) 54
$4x - 4 = z$
$4x - 4 = 54$
$4x - 4 + 4 = 54 + 4$
$4x = 58$
$\dfrac{4x}{4} = \dfrac{58}{4}$
$x = 14\dfrac{1}{2}$

58 is not divisible by 4

Now try (D) 60
$4x - 4 = z$
$4x - 4 = 60$
$4x - 4 + 4 = 60 + 4$
$4x = 64$
$\dfrac{4x}{4} = \dfrac{64}{4}$
$x = 16$
64 is divisible by 4 ✓

61

Chapter 5
Sneaky Plug In

There's one other type of plug in you need to look out for, and it's the sneakiest one of all. Let's review the types of plug ins.

	Basic Plug-In	A.C.T. "Answer Choice Test"	Sneaky Plug In
How to spot?	• Variables in the answer choices	• Numbers in the answer choices • The question asks for a single, specific thing	• Numbers in the answer choices, particularly fractions and percents • The question asks for a relationship
How to solve?	• Plug in our own numbers	• Plug in the answer choice numbers	• Plug in our own numbers

Remember the **Basic Plug In**? How do we spot it? There are variables in the answer choices! How do we solve it? Plug in our own numbers for the variables, work out arithmetically, box our answer, and then go through the answer choices with our numbers to find our boxed answer. Remember to try all of the answer choices.

How do we spot when we should use the **Answer Choice Test**? Numbers will be in the answer choices and the question will ask for a single, specific thing, such as "what is the value of y," or "how many shoes does Irene have?" And how do we solve? Backtrack the answer choice numbers into the question and see if the number works for every component of the problem. Remember to start with (C), unless the question asks for the least or greatest number.

And now on to our final Plug In – the **Sneaky Plug In**. How do we spot it? **Numbers in the answer choices, particularly fractions and percents,** (it looks just like an A.C.T. problem) but **the question asks for a relationship**. For instance, instead of "what is the value of x," the question will ask, "what is the ratio of x to y?" Instead of "how old is Bob," the question will ask, "how much older is Bob than Lou?" Whenever a question asks for a percent or a fractional part of something, you are on a Sneaky Plug In. How do we solve? **We plug in our own numbers and work out the problem arithmetically**. Let's do a few examples.

10. If r and s are integers such that $r > s$ and $r^2 + s^2 = 20$, which of the following can be the value of $r - s$?

  I. -2
  II. 2
  III. 6

(A) I only
(B) II only
(C) I and II only
(D) II and III only
(E) I, II, and III

Explanation:
It's tempting to treat this as an A.C.T. problem, but notice that the question does not ask for a single, specific thing. It asks for a relationship between r and s.

Let's throw in numbers for r and s that satisfy the restriction $r^2 + s^2 = 20$.
Say, $r = 4$, $s = 2$
$r^2 + s^2 = 20$
$4^2 + 2^2 = 20$
$16 + 4 = 20$ ✓

Plug in for $r - s$
$4 - 2 = 2$. We see that $r - s$ can be 2, which is Roman numeral II, so, eliminate the only answer choice that doesn't contain II, which is (A):

~~(A)~~ I only
(B) II only
(C) I and II only
(D) II and III only
(E) I, II, and III

It doesn't say anything about r and s being only positive integers, so let's throw in $r = 4$, $s = -2$. Check the restriction:
$r^2 + s^2 = 20$
$4^2 + (-2^2) = 20$
$16 + 4 = 20$ ✓

Plug in for $r - s$
$4 - (-2) = 6$
$4 + 2 = 6$.
So, $r - s$ can be 6. Eliminate all answer choices that don't contain III; these are (B) and (C).

~~(A)~~ I only
~~(B)~~ II only
~~(C)~~ I and II only
(D) II and III only
(E) I, II, and III

We have (D) and (E) left.
The only way to satisfy I is to make $r = 2$ and $s = 4$ ($2 - 4 = -2$) or $r = -4$ and $s = -2$ ($-4 - (-2) = -2$), but r has to be greater than s, so those aren't valid values for r and s.

~~(A)~~ I only
~~(B)~~ II only
~~(C)~~ I and II only
(D) II and III only
~~(E)~~ I, II, and III

Answer: (D) II and III only

Let's try a geometry problem.

16. If the base of a triangle is decreased by 20% and the height of the same triangle is increased by 20%, what is the effect on the area of the triangle?

(A) It is increased by 20%
(B) It is increased by 40%
(C) It is decreased by 4%
(D) It is decreased by 10%
(E) It is unchanged

Explanation:
Notice how there are percents in the answer choices and the question asks for a relationship. Let's plug in a base and height that will give us an area of 100, so that we don't have to worry about converting to percents later.

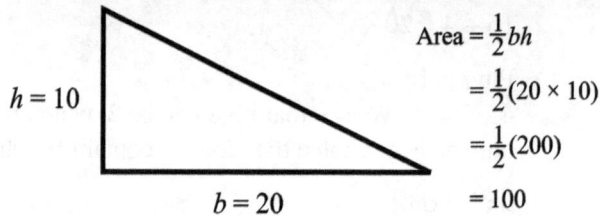

$h = 10$

$b = 20$

$$\text{Area} = \tfrac{1}{2}bh$$
$$= \tfrac{1}{2}(20 \times 10)$$
$$= \tfrac{1}{2}(200)$$
$$= 100$$

Decrease the base by 20%:
$20 - (20 \times .20) =$
$20 - 4 = 16$

Increase the height by 20 %:
$10 + (10 \times .20) =$
$10 + 2 = 12$

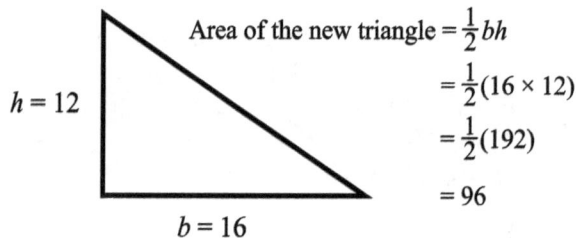

$h = 12$

$b = 16$

Area of the new triangle $= \tfrac{1}{2}bh$
$$= \tfrac{1}{2}(16 \times 12)$$
$$= \tfrac{1}{2}(192)$$
$$= 96$$

Area of original triangle – Area of new triangle = 100 – 96 = 4

4 is what percent of 100 (the area of the original triangle)? 4%! So, the triangle is decreased by 4%.

Answer: (C) It is decreased by 4%

___

Sneaky Plug in problems are all over the Grid Ins. Let's see how they work.

10. If 2/5 of the fish in Sam's fish tank are goldfish, 1/3 are guppies, and the remaining fish are minnows, what fraction of the fish in Sam's tank are minnows?

Explanation:
Notice that ETS asks for a relationship: *what fraction of the fish in the tank are minnows?*

Look at the denominators of the given fractions: 5 and 3. The Lowest Common Denominator is 15. So Plug In 15 as the total number of fish. Then just do the arithmetic. Remember that the word "of" always stands for multiplication:

$\tfrac{2}{5} \times 15 = 6$ goldfish $\tfrac{1}{3} \times 15 = 5$ guppies

Total goldfish and guppies = 6 + 5 = 11
15 total fish – 11 goldfish and guppies = 4 minnows.
Since it asks for a fraction, put the 4 minnows over the total for 4/15.

No matter what number we plug in for the total goldfish we will always get the reduced fraction 4/15.

Answer: 4/15

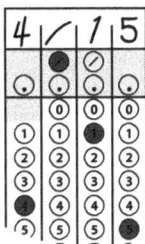

This next one is definitely a tricky one.

18. Roger rode his bike to school at an average speed of 15 miles per hour. He rode the bus home that afternoon along the same route. If the bus traveled at an average speed of 35 miles per hour and Roger spent a total of 40 minutes traveling to and from school, how many miles did Roger bike to school in the morning?

Explanation:
This is a weighted average problem so the key to making this a simple arithmetic problem is to throw in a common multiple of 15 and 35 for the distance. Let's plug in distance = 105 miles. Set up a proportion to find out how many hours it takes Roger to ride his bike a distance of 105 miles.

Bike to school:
$$\frac{15 \text{ miles}}{1 \text{ hour}} \times \frac{105 \text{ miles}}{x \text{ hours}}$$
$$15x = 105$$
$$\frac{15x}{15} = \frac{105}{15}$$
$$x = 7 \text{ hours}$$

Bus ride home:
$$\frac{35 \text{ miles}}{1 \text{ hour}} \times \frac{105 \text{ miles}}{y \text{ hours}}$$
$$35y = 105$$
$$\frac{35y}{35} = \frac{105}{35}$$
$$y = 3 \text{ hours}$$

Find the total travel time:
7 + 3 = 10 hours

Find the total roundtrip distance:

105 + 105 = 210 miles

Find Roger's rate:

$\frac{210 \text{ miles}}{10 \text{ hours}}$ = 21 miles per hour

Now we have to use the information that Roger spent 40 minutes traveling to and from school. We have our information in terms of hours (21 miles per hour), so we need to convert to minutes. There are 60 minutes in one hour, so put 21 miles over 60 minutes and set up a proportion to solve for the number of miles in 40 minutes.

$$\frac{21 \text{ miles}}{60 \text{ minutes}} = \frac{z \text{ miles}}{40 \text{ minutes}}$$

$$60z = 840$$

$$\frac{60z}{60} = \frac{840}{60}$$

$$z = 14 \text{ miles}$$

The question asks how many miles Roger biked to school in the morning. It's the same route to and from school so divide 14 by 2 and he biked a distance of 7 miles in the morning.

Answer: 7

Now try it on your own with this drill of six Sneaky Plug In questions.

Sneaky Plug In Drill

$$y = 5s$$
$$s = 6q$$
$$y = wq$$

11. According to the system of equations above, if $y \neq 0$, what is the value of w?

12. If $g(x)$ is defined by $g(x) = |5x - 13|$, what is one possible value of h for which $g(h) < h$?

13. The number k is a two-digit number. When k is divided by 7 the remainder is 6, and when k is divided by 6 the remainder is 5. What could be the value of k?

14. If $y > 0$, and y percent of 32 is equal to 64 percent of x, what is the value of $\frac{y}{x}$?

(A) $\frac{1}{3}$

(B) $\frac{2}{5}$

(C) $\frac{1}{2}$

(D) 3

(E) 2

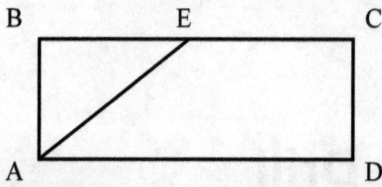

(Note: Figure not drawn to scale.)

14. In the figure, the area of rectangle ABCD is 12 and $\overline{BE} = \frac{4}{7}\overline{AD}$. What is the area of $\triangle ABE$?

(A) $\frac{35}{6}$

(B) $\frac{3}{5}$

(C) 1

(D) $\frac{24}{7}$

(E) $\frac{7}{3}$

18. If the area of circle P is four times that of circle M, then $\dfrac{\text{circumference of P}}{\text{circumference of M}} =$

(A) $\frac{1}{2}$

(B) 2

(C) 4

(D) 2π

(E) 4π

68

Answers and Explanations

Answer Key:

11. 30	**13.** 41 or 83	**14.** (D)
12. 3	**14.** (E)	**18.** (B)

$$y = 5s$$
$$s = 6q$$
$$y = wq$$

11. According to the system of equations above, if $y \neq 0$, what is the value of w?

Answer: 30

Explanation:
Plug in $s = 2$

Step 1:

$y = 5s$

$y = 5(2)$

$y = 10$

Step 2:

$s = 6q$

$2 = 6q$

$\dfrac{2}{6} = \dfrac{6q}{6}$

$q = \dfrac{2}{6} = \dfrac{1}{3}$

Step 3:

$y = wq$

$10 = w\dfrac{1}{3}$

$\dfrac{10}{1} \times \dfrac{w}{3}$

$w = 30$

---

12. If $g(x)$ is defined by $g(x) = |5x - 13|$, what is one possible value of h for which $g(h) < h$?

Answer: 3

Explanation:

Step 1:
Plug in and use trial and error:
Let $h = 3$. h is in the x spot,
so h is the same as x.

$g(x) = |5x - 13|$

$g(3) = |5(3) - 13|$

$g(3) = |15 - 13|$

$g(3) = |2|$

$g(3) = 2$

check $g(h) < h$

$\quad g(3) < 3$

$\quad g(3) = 2$

$\quad\quad 2 < 3$ ✓

So h could equal 3.

Step 2:
3 works; let's do one that may
not, so you can see the difference.
Let $x = 2$

$g(x) = |5x - 13|$

$g(2) = |5(2) - 13|$

$g(2) = |10 - 13|$

$g(2) = |-3|$

$g(2) = 3$

check $g(h) < h$

$\quad g(2) < 2$

$\quad g(2) = 3$

$\quad\quad 3 < 2$ ✗

---

13. The number k is a two-digit number. When k is divided by 7 the remainder is 6, and when k is divided by 6 the remainder is 5. What could be the value of k?

Answer: 41, 83

or

Explanation:

$\frac{k}{7} = ?$ R6

So start with a multiple of 7, such as 7, and then add 6 to it.

$7 + 6 = 13$

Plug in $k = 13$

$\frac{13}{6} = 2$ R1

We need a remainder of 5, not 1, so start with the next multiple of 7, which is 14, and add 6 to it.

$14 + 6 = 20$

Plug in $k = 20$

$\frac{20}{6} = 3$ R2

We have established a pattern. The first multiple of 7 yielded a remainder of 1, the second multiple of 7 a remainder of 2, so let's jump ahead to the 5th multiple of 7 to see if we can get a remainder of 5.

$5(7) + 6 =$

$35 + 6 = 41$

$\frac{41}{6} = 6$ R5

So 41 is one possibility for k. The pattern repeats itself, and 83 is a possibility as well.

---

14. If $y > 0$, and y percent of 32 is equal to 64 percent of x, what is the value of $\frac{y}{x}$?

(A) $\frac{1}{3}$

(B) $\frac{2}{5}$

(C) $\frac{1}{2}$

(D) 3

(E) 2

Explanation:

Notice the fractions in the answer choices and that the question itself is a percent problem. First, translate the question:

y percent of 32 is equal to 64 percent of x

$$\left(\frac{y}{100} \cdot 32 = \frac{64}{100} \cdot x \right)$$

70

Because it is a percent problem, plug in 100 for x and solve for y:

$$\frac{y}{100} \cdot 32 = \frac{64}{100} \cdot 100$$

$$\frac{32y}{100} = \frac{64}{100}_{1} \cdot \cancel{100}^{1} \qquad \text{Cross cancel.}$$

$$\frac{32y}{100} \diagup \frac{64}{1}$$

$$32y = 6400$$

$$\frac{32y}{32} = \frac{6400}{32}$$

$$y = 200$$

$$\frac{y}{x} = \frac{200}{100} = 2$$

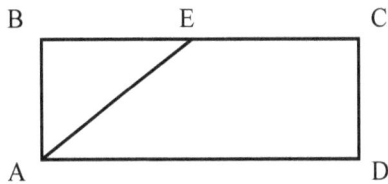

B   E   C

A   D

(Note: Figure not drawn to scale.)

14. In the figure, the area of rectangle ABCD is 12 and $\overline{BE} = \frac{4}{7}\overline{AD}$. What is the area of $\triangle ABE$?

(A) $\frac{35}{6}$

(B) $\frac{3}{5}$

(C) 1

(D) $\frac{24}{7}$ ⟵ circled

(E) $\frac{7}{3}$

Explanation:
Because it says "figure not drawn to scale," we can't make any assumptions.

Step 1:
Plug in a length and height for the rectangle that will give an area of 12, such as 2 and 6.
Let $\overline{AB} = 2$ and $\overline{AD} = 6$
So, $\overline{BE} = \frac{4}{7}(6) = \frac{24}{7}$

Step: 2:
Plug our values for AB and BE into the formula for area of a triangle.

Area of $\triangle ABE = \frac{1}{2}(\overline{AB} \cdot \overline{BE})$

$$= \frac{1}{2}\left(2 \cdot \frac{24}{7}\right)$$

Cross cancel: $= \frac{1}{2}_{1}\left(\frac{2^{1}}{1} \cdot \frac{24}{7}\right)$

$$= \frac{24}{7}$$

18. If the area of circle P is four times that of circle M, then $\dfrac{\text{circumference of P}}{\text{circumference of M}} =$

(A) $\dfrac{1}{2}$

(B) 2

(C) 4

(D) 2π

(E) 4π

Explanation:

Plug in 4π for the area of circle M and solve for the radius of each circle. Area $= \pi r^2$

Circle M

$4\pi = \pi r^2$

$\dfrac{4\pi}{\pi} = \dfrac{\pi r^2}{\pi}$

$4 = r^2$

$\sqrt{4} = \sqrt{r^2}$

$r = 2$

circumference $= 2\pi r$

$c = 2\pi \cdot 2$

$c = 4\pi$

Circle P

$4\pi \times 4 = 16\pi$

$16\pi = \pi r^2$

$\dfrac{16\pi}{\pi} = \dfrac{\pi r^2}{\pi}$

$16 = r^2$

$\sqrt{16} = \sqrt{r^2}$

$r = 4$

circumference $= 2\pi r$

$c = 2\pi \cdot 4$

$c = 8\pi$

$\dfrac{\text{circumference of P}}{\text{circumference of M}} = \dfrac{8\pi}{4\pi} = 2$

Chapter 6
Comprehensive Plug In Drill

5. Of the students enrolled in Evergreen College's MBA program, two-thirds have experience in the work force; of those, two-thirds have five or more years of experience in the work force. Which of the following circle graphs could represent the MBA students, divided into three different groups: those with five or more years experience in the work force, those with less than five years of work experience, and those with no work experience.

(A)

(B)

(C)

(D)

(E)

6. If f and g are positive numbers and $f - g = 11$, then $\dfrac{11 + g}{f} =$

(A) $g - 1$
(B) f
(C) 1
(D) 0
(E) −1

8. If S and R are two sets of integers, and if every even number in set S is also in set R, all of the following could be true EXCEPT

(A) 3 is in neither S nor R.
(B) 4 is in both S and R.
(C) 5 is in R, but not in S.
(D) 6 is in S, but not in R.
(E) If 10 is not in S, then 10 is not in R.

9. If $\dfrac{2r - 2s}{p + q} = \dfrac{2}{5}$, then $\dfrac{5r - 5s}{6p + 6q} =$

(A) $\dfrac{1}{6}$
(B) $\dfrac{1}{3}$
(C) $\dfrac{2}{5}$
(D) $\dfrac{3}{5}$
(E) $\dfrac{5}{6}$

9. The tick marks on the number line shown above are equally spaced. What is the value of z?

(A) 5
(B) 7
(C) 10
(D) 12
(E) 32

10. If q is an integer, which of the following is NOT equal to $(2^6)^q$?

(A) 8^{2q}

(B) 4^{3q}

(C) $2^q(2^{5q})$

(D) 64^q

(E) $4^q(2^{2q})$

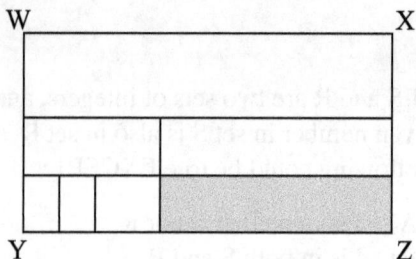

12. In the figure above, rectangle WXYZ is made up of 7 non-overlapping rectangles. The two smallest rectangles have the same area. Each of the other rectangles has twice the area of the next smaller rectangle. The shaded rectangle is what fraction of the area of WXYZ?

(A) $\frac{1}{10}$

(B) $\frac{1}{5}$

(C) $\frac{2}{5}$

(D) $\frac{1}{16}$

(E) $\frac{1}{2}$

13. If $f(x) > 1$ for all real values of x, which of the following equations could be the function of f?

(A) $f(x) = x + 2$

(B) $f(x) = x - 2$

(C) $f(x) = x^2 + 2$

(D) $f(x) = x^2 - 2$

(E) $f(x) = x^3 + 1$

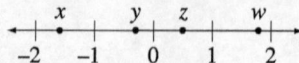

15. If x, y, z and w represent the coordinates of the points on the number line above, which of the following is the least?

(A) $|x + w|$

(B) $|x - w|$

(C) $|x + z|$

(D) $|x - z|$

(E) $|x + y|$

15. A group of people is to equally share the cost of a car represented by x dollars. In terms of x, how many dollars less will each person contribute if there are 5 instead of 3 people in the group?

(A) $\frac{x}{15}$

(B) $\frac{x}{5}$

(C) $\frac{x}{3}$

(D) $\frac{2x}{15}$

(E) $\frac{8x}{15}$

15. If $a^2 = b^4$ and $b > 1$, then in terms of b, what does $a^{1/2}$ equal?

(A) $\frac{b^6}{3}$

(B) $\frac{b^5}{2}$

(C) $\frac{b^4}{2}$

(D) b^2

(E) b

16. If $k \geq 3$, how much less than $5k - 2$ is $3k + 1$ in terms of k?

(A) $2k + 5$

(B) $k - 5$

(C) $k + 5$

(D) $2k - 3$

(E) $k - 1$

Tony's Monthly Expenses

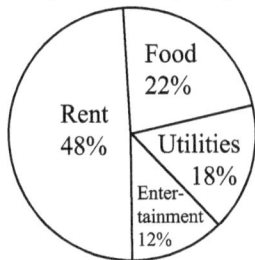

16. Tony's monthly living expenses are shown in the graph above. Tony's monthly income is $1,500 and he shares an apartment with two other people. If the rent is shared equally, what is the total cost of the apartment?

(A) 720
(B) 1440
(C) 1500
(D) 2160
(E) 2230

x	−2	0	2
$g(x)$	4	16	64

17. Some values for the function g are shown in the table. If $g(x) = c^2 d^x$, and c and d are positive constants, what is the value of d?

(A) 1/2
(B) 2
(C) 4
(D) 8
(E) 16

18. The sum of two consecutive odd integers is x. What is the sum of the next two consecutive odd integers greater than x?

(A) $x + 5$
(B) $x + 8$
(C) $2x + 3$
(D) $2x + 4$
(E) $2x + 5$

19. Let $\blacksquare k$ be defined as $\blacksquare k = k + \frac{3}{k}$. If k is a non-zero integer, and $\blacksquare k = 2n$, what is a possible value of n?

(A) −3
(B) −2
(C) −1
(D) 0
(E) 1

Answers and Explanations

5. Of the students enrolled in Evergreen College's MBA program, two-thirds have experience in the work force; of those, two-thirds have five or more years of experience in the work force. Which of the following circle graphs could represent the MBA students, divided into three different groups: those with five or more years experience in the work force, those with less than five years of work experience, and those with no work experience.

(A)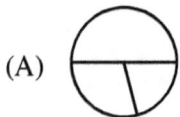

(B)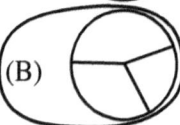

(C)

(D)

(E)

Explanation:
Let's turn this Sneaky Plug In into a simple arithmetic problem by plugging in a multiple of 3 for the number of MBA students. Why? Because 3 is in the denominator of the given fractions. Let's use 90.

$90 \times \frac{2}{3} = \frac{180}{3} = 60$ students have work experience

$60 \times \frac{2}{3} = \frac{120}{3} = 40$ students with 5 or more years work experience

$60 - 40 = 20$ students with less than 5 years work experience

$90 - 60 = 30$ students with no work experience

If the pie chart represents 90 total, then 40 is less than half of 90, so we can eliminate (A), (D) and (E). (C) divides the graph equally into thirds, so (B) best represents the 40, 20, 30 breakdown.

---

6. If f and g are positive numbers and $f - g = 11$, then $\frac{11+g}{f} =$

(A) $g - 1$
(B) f
(C) 1
(D) 0
(E) -1

Explanation:

Notice the variables in the answer choices and solve as a Basic Plug In.

Plug in $f = 19$ and $g = 8$.

Restriction: $f - g = 11$

$19 - 8 = 11$ Matches restriction. ✓

$$\frac{11 + g}{f} =$$

$$\frac{11 + 8}{19} = \frac{19}{19} = 1 \quad \boxed{1}$$

Plug in f and g to the answer choices:

(A) $g - 1$
       $8 - 1 = 7$

(B) f
       19

(C) 1 ←

(D) 0

(E) -1

---

8.   If S and R are two sets of integers, and if every even number in set S is also in set R, all of the following could be true EXCEPT

(A) 3 is in neither S nor R.
(B) 4 is in both S and R.
(C) 5 is in R, but not in S.
(D) 6 is in S, but not in R.
(E) If 10 is not in S, then 10 is not in R.

Explanation:

The word EXCEPT tells us that we are looking to cancel possibly true statements. We are looking for a FALSE statement on this Sneaky Plug In.

Suppose S and R are as follows:

S: $\{1, 2, 3, 4, 5\}$

R: $\{2, 4, 5\}$

This meets our restriction, since every even member of S (2, 4) is also in R. Go through the answer choices and apply them to the sets S and R.

(A) 3 is in neither S nor R.

3 happens to be in our current S, but no stated restriction makes it necessary that 3 is in S, so we can change S to exclude 3:

S: $\{1, 2, 4, 5\}$

R: $\{2, 4, 5\}$

(B) 4 is in both S and R.

You are not trying to disprove this rule. The question asks for "could be true," which holds. Since 4 is even and could be included in sets S and R, (B) could be true and is thus canceled.

S: $\{1, 2, 5\}$

R: $\{2, 4, 5\}$

(C) 5 is in R, but not in S.
The restriction says nothing about odd numbers. So we can take 5 out of S without violating the restriction. This could be true, but we want a false statement. Cancel.
S: {1, 2}
R: {2, 4, 5}

✓ (D) 6 is in S, but not in R.
The restriction says that all even numbers in S must be in R, and 6 is even, so it's not possible to have 6 in S without 6 being in R. This is false!

(E) If 10 is not in S, then 10 is not in R.
This is consistent with the restriction as well as our chosen sets of S and R. 10 is not in S and also not in R.

---

9. If $\dfrac{2r-2s}{p+q} = \dfrac{2}{5}$, then $\dfrac{5r-5s}{6p+6q} =$

(A) $\dfrac{1}{6}$

(B) $\dfrac{1}{3}$

(C) $\dfrac{2}{5}$

(D) $\dfrac{3}{5}$

(E) $\dfrac{5}{6}$

Explanation:
Here's another Sneaky Plug In.
Plug in numbers for $r, s, p,$ and q that make the expression $\dfrac{2r-2s}{p+q} = \dfrac{2}{5}$ true.

Let $r = 4, s = 3, p = 3, q = 2$:

$$\dfrac{2r-2s}{p+q} =$$

$$\dfrac{2(4)-2(3)}{3+2} =$$

$$\dfrac{8-6}{5} = \dfrac{2}{5}$$

Plug these numbers into the second equation: $\dfrac{5r-5s}{6p+6q} =$

$$\dfrac{5(4)-5(3)}{6(3)+6(2)} =$$

$$\dfrac{20-15}{18+12} = \dfrac{5}{30} = \dfrac{1}{6}$$

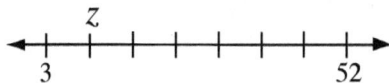

9. The tick marks on the number line shown above are equally spaced. What is the value of z?

(A) 5
(B) 7
(C) 10
(D) 12
(E) 32

Explanation:
Notice that there are numbers in the answer choices and the question asks for a single, specific thing: z. Use the A.C.T. Plug in (C) 10:
$z = 10$, and the difference between 3 and 10 is 7:

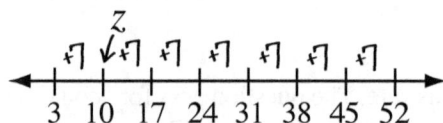

There is a distance of 7 between each of our 7 "equally spaced" tick marks. $7 \times 7 = 49$
Does 49 match the distance between 52 and 3? $52 - 3 = 49$
So it works. No need to plug in further.

78

10. If q is an integer, which of the following is NOT equal to $(2^6)^q$?

(A) 8^{2q}
(B) 4^{3q}
(C) $2^q(2^{5q})$
(D) 64^q
(E) $4^q(2^{2q})$

Explanation:

This is a "NOT" problem, so we are looking for a false answer on this Basic Plug In.

Let $q = 2$

So $(2^6)^q = (2^6)^2$

$= 2^{12} = 4096$ $\boxed{4096}$

Plug In $q = 2$ into the answer choices to find which answer choice does NOT match our boxed answer:

(A) 8^{2q}
$8^{(2 \cdot 2)}$
$8^4 = 4096$

(B) 4^{3q}
$4^{(3 \cdot 2)}$
$4^6 = 4096$

(C) $2^q(2^{5q})$
$2^2(2^{(5 \cdot 2)})$
$4(2^{10}) = 4096$

(D) 64^q
$64^2 = 4096$

(E) $4^q(2^{2q})$
$4^2(2^{(2 \cdot 2)})$
$16(2^4) = 256$

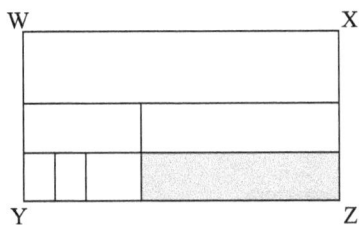

12. In the figure above, rectangle WXYZ is made up of 7 non-overlapping rectangles. The two smallest rectangles have the same area. Each of the other rectangles has twice the area of the next smaller rectangle. The shaded rectangle is what fraction of the area of WXYZ?

(A) $\frac{1}{10}$

(B) $\frac{1}{5}$

(C) $\frac{2}{5}$

(D) $\frac{1}{16}$

(E) $\frac{1}{2}$

Explanation:

This is a Sneaky Plug In. Plug in 2 for the area of the 2 smallest rectangles. The next largest rectangle has an area of 4. The area of the next largest is 8. The area of the next two is 16, and the area of the largest is 32.

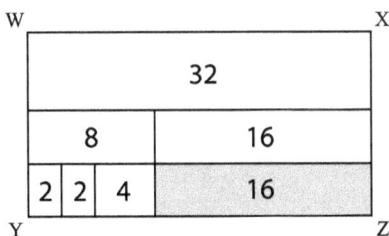

Total area $= 2 + 2 + 4 + 8 + 16 + 16 + 32 = 80$

The question is asking for a fraction, so put the shaded area over the total area: $\frac{16}{80} = \frac{1}{5}$

13. If $f(x) > 1$ for all real values of x, which of the following equations could be the function of f?

(A) $f(x) = x + 2$
(B) $f(x) = x - 2$
(C) $f(x) = x^2 + 2$
(D) $f(x) = x^2 - 2$
(E) $f(x) = x^3 + 1$

Explanation:
This is a Basic Plug In. Plug in $x = 0$:

(A) $f(x) = x + 2$
$f(0) = 0 + 2$
$= 2$
$2 > 1$ ✓

(D) $f(x) = x^2 - 2$
$f(0) = 0^2 - 2$
$= -2$
$-2 \not> 1$

(B) $f(x) = x - 2$
$f(0) = 0 - 2$
$= -2$
$-2 \not> 1$

(E) $f(x) = x^3 + 1$
$f(0) = 0^3 + 1$
$= 1$
$1 \not> 1$

(C) $f(x) = x^2 + 2$
$f(0) = 0^2 + 2$
$= 2$
$2 > 1$ ✓

Eliminate (B), (D), and (E). Now try plugging in $x = -3$:

(A) $f(x) = x + 2$
$f(-3) = -3 + 2$
$= -1$
$-1 \not> 1$

(C) $f(x) = x^2 + 2$
$f(-3) = -3^2 + 2$
$= 11$
$11 > 1$ ✓

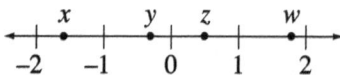

15. If x, y, z and w represent the coordinates of the points on the number line above, which of the following is the least?

(A) $|x + w|$
(B) $|x - w|$
(C) $|x + z|$
(D) $|x - z|$
(E) $|x + y|$

Explanation:
This is a Basic Plug In. Let $x = -1.6$, $y = -.3$, $z = .5$ and $w = 1.8$. Plug in to find the expression with the lowest value:

(A) $|x + w|$
$|-1.6 + 1.8| = |.2| = .2$

(D) $|x - z|$
$|-1.6 - .5| = |-2.1| = 2.1$

(B) $|x - w|$
$|-1.6 - 1.8| = |-3.4| = 3.4$

(E) $|x + y|$
$|-1.6 + (-.3)| = |-1.9| = 1.9$

(C) $|x + z|$
$|-1.6 + .5| = |-1.1| = 1.1$

.2 is the least value, so the answer is (A) $|x + w|$

80

15. A group of people is to equally share the cost of a car represented by x dollars. In terms of x, how many dollars less will each person contribute if there are 5 instead of 3 people in the group?

(A) $\frac{x}{15}$

(B) $\frac{x}{5}$

(C) $\frac{x}{3}$

(D) $\frac{2x}{15}$

(E) $\frac{8x}{15}$

Explanation:
This is a Basic Plug In. Let $x = 150$

If there are 5 people, then $\frac{150}{5} = 30$

If there are 3 people, then $\frac{150}{3} = 50$

$\boxed{20}$

and $50 - 30 = 20$

Plug in 150 to the answer choices:

(A) $\frac{x}{15}$

$\frac{150}{15} = 10$

(D) $\frac{2x}{15}$

$\frac{2(150)}{15} = \frac{300}{15} = 20$

(B) $\frac{x}{5}$

$\frac{150}{5} = 30$

(E) $\frac{8x}{15}$

$\frac{8(150)}{15} = \frac{1200}{15} = 80$

(C) $\frac{x}{3}$

$\frac{150}{3} = 50$

---

15 If $a^2 = b^4$ and $b > 1$, then in terms of b, what does $a^{\frac{1}{2}}$ equal?

(A) $\frac{b^6}{3}$

(B) $\frac{b^5}{2}$

(C) $\frac{b^4}{2}$

(D) b^2

(E) b

Explanation:
This is a Basic Plug In. Let $b = 2$.
$a^2 = b^4$
$a^2 = 2^4$
$a^2 = 16$
Solve for a by taking the square root of both sides:
$\sqrt{a^2} = \sqrt{16}$
$a = 4$
So, $a^{\frac{1}{2}} = 4^{\frac{1}{2}}$.
When a number is raised to a fraction, the numerator stands for the power and the denominator stands for the square root.
$4^{\frac{1}{2}} = \sqrt[2]{4^1}$ or simply $\sqrt{4} = 2$.

$\boxed{2}$

Plug $b = 2$ into the answer choices:

(A) $\dfrac{b^6}{3}$

$\quad \dfrac{2^6}{3} = \dfrac{64}{3}$

(D) b^2

$\quad 2^2 = 4$

(B) $\dfrac{b^5}{2}$

$\quad \dfrac{2^5}{2} = \dfrac{32}{2} = 16$

(E) b

$\quad 2$

(C) $\dfrac{b^4}{2}$

$\quad \dfrac{2^4}{2} = \dfrac{16}{2} = 8$

---

16. If $k \geq 3$, how much less than $5k - 2$ is $3k + 1$ in terms of k?

(A) $2k + 5$
(B) $k - 5$
(C) $k + 5$
(D) $2k - 3$
(E) $k - 1$

Explanation:
This is a Basic Plug In. Let $k = 4$.
$5k - 2 =$
$5(4) - 2 =$
$20 - 2 = 18$

$3k + 1 =$
$3(4) + 1 =$
$12 + 1 = 13$

"how much less than $5k - 2$ is $3k + 1$"
$18 - 13 = 5$ $\quad \boxed{5}$

Plug $k = 4$ in to the answer choices:

(A) $2k + 5$
$\quad 2(4) + 5$
$\quad 8 + 5 = 13$

(D) $2k - 3$
$\quad 2(4) - 3$
$\quad 8 - 3 = 5$

(B) $k - 5$
$\quad 4 - 5 = -1$

(E) $k - 1$
$\quad 4 - 1 = 3$

(C) $k + 5$
$\quad 4 + 5 = 9$

Tony's Monthly Expenses

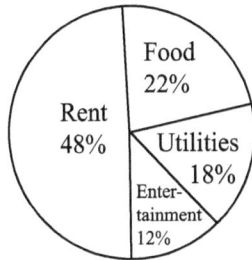

16. Tony's monthly living expenses are shown in the graph above. Tony's monthly income is $1,500 and he shares an apartment with two other people. If the rent is shared equally, what is the total cost of the apartment?

(A) 720
(B) 1440
(C) 1500
(D) 2160
(E) 2230

Explanation:
Use the A.C.T.
Plug in (C) 1500
Be sure to divide 1500 by 3, not 2, because there are 3 people in the apartment: Tony and 2 roommates.

$$\frac{1500}{3} = \$500$$

Tony spends 48% of his monthly income on rent, so:

$$500 = \frac{48}{100}(1500)$$

$$500 \neq 720$$

Now try (D) 2160

$$\frac{2160}{3} = 720$$

$$720 = \frac{48}{100}(1500)$$

$$720 = 720 \quad \checkmark$$

x	-2	0	2
$g(x)$	4	16	64

17. Some values for the function g are shown in the table. If $g(x) = c^2 d^x$, and c and d are positive constants, what is the value of d?

(A) 1/2
(B) 2
(C) 4
(D) 8
(E) 16

Explanation:
To solve this question it is important to know that any number raised to an exponent of 0 is 1. For example, $6^0 = 1$. Notice how they give you an x value of 0 in the table. Use it!

Use the second value for $g(x)$ to solve for the constant c. c and d are **constants**, which means they are the same number, regardless of the x and y values.

$x = 0, g(x) = 16$

$g(x) = c^2 d^x$

$16 = c^2 d^0$

$16 = c^2 1$

$\sqrt{16} = \sqrt{c^2}$

$c = 4$

Use the A.C.T. Plug in (C) 4 for d.

Use the first value in the table for $g(x)$, $x = -2, c = 4$

$4 = c^2 d^x$

$4 = 4^2 \bullet 4^{-2}$

$4 = 16 \bullet 4^{-2}$ $\leftarrow$ (When an integer is raised to a negative exponent, put the integer under 1 and make the exponent positive!

$4 = 16 \bullet \dfrac{1}{4^2}$

$4 = 16 \bullet \dfrac{1}{16}$

$4 = \dfrac{16}{16}$

$4 \neq 1$

Try (B) 2:

$4 = c^2 d^x$

$4 = 4^2 \bullet 2^{-2}$

$4 = 16 \bullet 2^{-2}$

$4 = 16 \bullet \dfrac{1}{2^2}$

$4 = 16 \bullet \dfrac{1}{4}$

$4 = \dfrac{16}{4}$

$4 = 4$ ✓

---

18. The sum of two consecutive odd integers is x. What is the sum of the next two consecutive odd integers greater than x?

(A) $x + 5$
(B) $x + 8$
(C) $2x + 3$
(D) $2x + 4$
(E) $2x + 5$

Explanation:
This is a Basic Plug In. Plug in 3 and 5 for the consecutive odd integers.
So $x = 3 + 5 = 8$
$x = 8$.

The next two consecutive odd integers greater than 8 are 9 and 11. $9 + 11 = 20$.
Box 20 and plug $x = 8$ into the answer choices: **20**

(A) $x + 5$
  $8 + 5 = 13$

(B) $x + 8$
  $8 + 8 = 16$

(C) $2x + 3$
  $2(8) + 3 = 19$

(D) $2x + 4$
  $2(8) + 4 = 20$ $\leftarrow$

(E) $2x + 5$
  $2(8) + 5 = 21$

84

19. Let $\blacksquare k$ be defined as $\blacksquare k = k + \frac{3}{k}$. If k is a nonzero integer, and $\blacksquare k = 2n$, what is a possible value of n?

(A) -3
(B) -2
(C) -1
(D) 0
(E) 1

Explanation:
Do the A.C.T. Plug in (C) -1 for n

$$\blacksquare k = 2n$$

$$\blacksquare k = 2(-1)$$

$$\blacksquare k = -2$$

Since the function of k is equal to -2, use the given definition and set the entire function of k equal to -2 in order to solve for k.

$$\blacksquare k = k + \frac{3}{k}$$

$$-2 = k + \frac{3}{k}$$

$$\left[-2 = k + \frac{3}{k} \right] k$$

$$-2k = k^2 + 3$$

$$-2k + 2k = k^2 + 3 + 2k$$

$$0 = k^2 + 2k + 3$$

This equation cannot be factored. Try (B) -2

$$\blacksquare k = 2n$$

$$\blacksquare k = 2(-2)$$

$$\blacksquare k = -4$$

$$\blacksquare k = k + \frac{3}{k}$$

$$-4 = k + \frac{3}{k}$$

$$\left[-4 = k + \frac{3}{k} \right] k$$

$$-4k = k^2 + 3$$

$$-4k + 4k = k^2 + 3 + 4k$$

$$0 = k^2 + 4k + 3$$

$$0 = (k+1)(k+3)$$

This is the only equation that can be factored.

Chapter 7
Arithmetic Lesson

PEMDAS

One of the first things in arithmetic that trips students up is ***Order of Operations***. Let's use the acronym PEMDAS to help us remember the correct order in which to solve arithmetic.

P - parentheses
E - exponents
M - multiplication
D - division
A - addition
S – subtraction

Parentheses come first, followed by *Exponents*. Nothing confusing there. The tricky part comes with *Multiplication* and *Division*. Multiplication doesn't necessarily come first. The rule for multiplication/division is *do whatever comes first from left to right*. If division comes first, then make sure you divide before you multiply.

Take a look at these two problems:

$$9 \div 3 \times 2(4 - 1)^2$$

Parentheses first:	$9 \div 3 \times 2(3)^2$
Now Exponents:	$9 \div 3 \times 2(9)$
Multiplication/Division left to right (be sure to divide 9 by 3 first!):	$3 \times 2(9)$
And then multiply the rest across:	$3 \times 2 \times 9 = 54$

$$9 \times 9 \div 3(4 - 1)^2$$

Parentheses First:	$9 \times 9 \div 3(3)^2$
Now Exponents:	$9 \times 9 \div 3(9)$
Multiplication/Division left to right (be sure to multiply 9×9 first!):	$81 \div 3(9)$
Now divide 81 by 3 before you multiply by 9:	$27 \times 9 = 243$

Addition and *Subtraction* work the same way. Do whatever comes first left to right.

Try these on your own:

$$(3 + 2)^2 + 4 \times 3 - 2 + 1$$

Parentheses first:	$(5)^2 + 4 \times 3 - 2 + 1$
Now Exponents:	$25 + 4 \times 3 - 2 + 1$
Multiplication/Division left to right	$25 + 12 - 2 + 1$
Now, Addition/Subtraction left to right:	$37 - 2 + 1$
	$35 + 1$
	36

$$6 \times 9 + 3 - 2$$

Multiplication first:	$54 + 3 - 2$
Now, Addition:	$57 - 2$
Finally, Subtraction:	55

Remember: map out the problem on your test booklet first to avoid making a careless calculator mistake. You can either enter each step into your calculator separately, or plug the entire expression into your calculator. Just be sure to punch the numbers in correctly and include parentheses and exponents, and your calculator will do the work for you!

FRACTIONS

Unfortunately ETS probably won't be asking you to add, subtract, multiply, or divide fractions, as that would be a little too easy. And for the most part you can use your calculator for these mathematical functions anyway.

If you are going to work with fractions in your calculator, make sure you USE PARENTHESES around the fractions. Otherwise the calculator will not give you the right value. Keep in mind, however, that students are prone to errors when entering multiple actions at once.

You should also know how to convert from fraction to decimal and decimal to fraction on your calculator. On many TI calculators it's as simple as hitting **(MATH) (ENTER) (ENTER)**

Decimals work well for some students, but the fractions used on the SAT are usually pretty simple. Do the work on the page and then use your little robot friend to find computations.

Let's review some basic fraction rules.

MIXED AND IMPROPER FRACTIONS

$2\frac{2}{3}$ is a mixed fraction.

To convert to an improper fraction, multiply the whole number and denominator and add that product to the numerator, putting this value over the original denominator.

So: $2\frac{2}{3} = 2\frac{2}{3} = \frac{8}{3}$

Remember, you cannot grid in a mixed number. Either convert it to an improper fraction or a decimal.

ADDING AND SUBTRACTING FRACTIONS

When adding and subtracting fractions, find a common denominator, add or subtract the numerators, and put that sum over your denominator.

For example: $\frac{4}{7} + \frac{2}{3}$

$$\frac{4 \times 3}{7 \times 3} = \frac{12}{21}$$

$$+ \frac{2 \times 7}{3 \times 7} = \frac{14}{21}$$

$$\frac{26}{21}$$

MULTIPLYING FRACTIONS

Multiply across: numerator × numerator and denominator × denominator.

For example: $\frac{4}{5} \times \frac{10}{12} = \frac{40}{60} = \frac{2}{3}$

cross-canceling:

$$\frac{^1\cancel{4}}{_1\cancel{5}} \times \frac{\cancel{10}^2}{\cancel{12}_3} = \frac{2}{3}$$

DIVIDING FRACTIONS

Keep it, Switch it, Flip it. Keep the first fraction as is, *Switch* the division sign to a multiplication sign, and *Flip* the second fraction.

For example: $\frac{6}{7} \div \frac{2}{21} = \frac{6}{7} \times \frac{21}{2} = \frac{^3\cancel{6}}{_1\cancel{7}} \times \frac{\cancel{21}^3}{\cancel{2}_1} = \frac{3}{1} \times \frac{3}{1} = 9$

Let's talk *cross-canceling*. We cross-canceled on the previous multiplication and division problems. What if there is an equal sign between the two fractions? Can we cross cancel then? NO! Cross canceling is only allowed when multiplying. When there is an equal sign we can only *cross-multiply*.

For example: $\frac{2x}{6} = \frac{6}{10}$

$$\frac{2x}{6} \diagdown\hspace{-0.9em}\diagup \frac{6}{10}$$

$$20x = 36$$

$$\frac{20x}{20} = \frac{36}{20}$$

$$x = \frac{36}{20} = \frac{9}{5}$$

COMPARING FRACTIONS

When comparing fractions it is easy to convert them to decimals using your calculator. From there draw a number line and plot the decimals if you are still confused.

Here is another way to compare fractions: $\frac{11}{13}$ versus $\frac{7}{9}$

$$\frac{11}{13} \diagdown\hspace{-0.9em}\diagup \frac{7}{9}$$

$$11 \times 9 = 99 \quad 7 \times 13 = 91$$

$99 > 91$, so $\frac{11}{13}$ is the bigger fraction.

MEAN/MEDIAN/MODE

MEDIAN
Definition: the middle value of a set when the numbers are arranged from least to greatest

MODE
Definition: the number that occurs most frequently

We've discussed Median and Mode in the Math Vocabulary Chapter, so let's talk more in depth about Average.

AVERAGE

$$\text{Average (mean)} = \frac{\text{Sum}}{\text{\# of things}}$$

Typically it's not as easy as "what's the average?" Rather, ETS will give you the average and the number of things and you need to find the sum.

The Key to Average Problems: Multiply the average and the number of things to find the sum!

So if ETS says, "the average test scores of 5 students is 80" multiply 80×5 for a sum of 400. Now you're on your way to solving the rest of the problem.

Let's do some:

$$13 - r, 13, 13 + r, r^2$$

5. What is the average (arithmetic mean) of the 4 terms in the list above?

(A) 3
(B) 12
(C) 13
(D) $13 + \frac{r}{4}$
(E) $\frac{39 + r^2}{4}$

Explanation:
Notice the variables in the answer choices! Let's use our Basic Plug In technique.
Let $r = 2$
$13 - r, 13, 13 + r, r^2$
$13 - 2, 13, 13 + 2, 2^2$
$11, 13, 15, 4$
So the list contains $11, 13, 15, 4$

$$\boxed{\frac{43}{4}}$$

$$\frac{11 + 13 + 15 + 4}{4} = \frac{43}{4}$$

Plug $r = 2$ into the answer choices to find a match for your boxed answer:

(A) 3
(B) 12
(C) 13
(D) $13 + \frac{r}{4}$

$$13 + \frac{2}{4} = 13 + \frac{1}{2} = 13\frac{1}{2}$$

(E) $\frac{39 + r^2}{4}$

$$\frac{39 + 2^2}{4} = \frac{39 + 4}{4} = \frac{43}{4}$$

Answer: (E) $\frac{39 + r^2}{4}$

5. The average (arithmetic mean) of a and b is 12. The average of c and d is 14. What is the average of $a, b, c,$ and d?

(A) 6.5
(B) 13
(C) 17.5
(D) 26
(E) 52

Explanation:

Remember: when given the average the first step is to find the sum.

Step 1:

$$\frac{a+b}{2} = 12$$

$$\frac{a+b}{2} = \frac{12}{1}$$

cross-multiply:

$(a+b)\,1 = 12\,(2)$

$a + b = 24$

Step 2:

$$\frac{c+d}{2} = 14$$

$$\frac{c+d}{2} = \frac{14}{1}$$

cross-multiply:

$(c+d)\,1 = 14\,(2)$

$c + d = 28$

Step 3:

$$\frac{a+b+c+d}{4} =$$

Substitute 24 for $a + b$ and 28 for $c + d$.

$$\frac{24+28}{4} =$$

$$\frac{52}{4} = 13$$

Answer: (B) 13

PERCENTS

Percent translates to "out of 100." Any percent can be written as a fraction with 100 as the denominator. Fractions are the safest way to work with percents.

$$30\% = \frac{30}{100} \text{ or } .3 \qquad 3\% = \frac{3}{100} \text{ or } .03 \qquad .3\% = \frac{.3}{100} \text{ or } .003$$

Percent Problems can typically be solved by translating words into a mathematical equation. Here are the key terms to know:

ENGLISH	MATH
Percent	Divide by 100
What	A variable: x, n
Of	Multiply: ×
Is, are, was, were, equals	=
What Percent	$\frac{x}{100}$

Pay especially close attention to the "of" which means multiply, and the "what percent" which should always be written as $\frac{x}{100}$.

Let's do some translations for practice:

1. 33 percent of what is 30 percent of 55 percent of 400?

$$\frac{33}{100} \cdot x = \frac{30}{100} \cdot \frac{55}{100} \cdot 400$$

$$\frac{33}{100} x = \frac{660,000}{10,000}$$

Simplify: $\dfrac{33x}{100} = \dfrac{66\cancel{0,000}}{1\cancel{0,000}}$

90

$$\frac{33x}{100} \diagup\!\!\!\!\times \frac{66}{1}$$

$$33x = 6{,}600$$

$$\frac{33x}{33} = \frac{6{,}600}{33}$$

$$x = 200$$

2. 109 is what percent of 872?

$$109 = \frac{x}{100} \cdot 872$$

$$\frac{109}{1} \diagup\!\!\!\!\times \frac{872x}{100}$$

$$872x = 10{,}900$$

$$\frac{872x}{872} = \frac{10{,}900}{872}$$

$$x = 12.5$$

3. What percent of 49 is 12?

$$\frac{x}{100} \cdot 49 = 12$$

$$\frac{49x}{100} \diagup\!\!\!\!\times \frac{12}{1}$$

$$49x = 1200$$

$$\frac{49x}{49} = \frac{1200}{49}$$

$$x = 24.49$$

Let's try some SAT problems:

14. Which of the following is equivalent to $\frac{1}{3}$ of 29 percent of 813?

(A) $\frac{29}{3} \times 812$

(B) $\frac{29}{3}\%$ of 271

(C) $\frac{281}{3}\%$ of $\frac{271}{3}$

(D) 29% of $\frac{271}{3}$

(E) 29% of 271

Explanation:
Work it out as a translation problem:

$\frac{1}{3}$ of 29 percent of 813

$$\frac{1}{3} \times \frac{29}{100} \times 813 = 78.59$$

Box your value and go through the answer choices to find a match:

$$\boxed{78.59}$$

(A) $\frac{29}{3} \times 812 = 7849.3\overline{3}$

(B) $\frac{29}{3}\%$ of $271 = \frac{\frac{29}{3}}{100} \times 271 = \frac{29}{3} \times \frac{1}{100} \times 271 = 26.196\overline{6}$

(C) $\frac{281}{3}\%$ of $\frac{271}{3} = \frac{\frac{281}{3}}{100} \times \frac{271}{3} = \frac{281}{3} \times \frac{1}{100} \times \frac{271}{3} = 84.612\overline{2}$

(D) 29% of $\frac{271}{3} = \frac{29}{100} \times \frac{271}{3} = 26.196\overline{6}$

(E) 29% of $271 = \frac{29}{100} \times 271 = 78.59$ ⟵

Answer: (E) 29% of 271

---

15. Taj and Ron both work at the same furniture store. Taj's weekly salary consists of $450 plus 30 percent of her sales. Ron's weekly salary consists of $600 plus 15 percent of his weekly sales. If in the second week of March they both had the same amount of sales and the same weekly salary, what was the weekly salary in dollars?

(A) 480
(B) 550
(C) 750
(D) 580
(E) 650

Explanation:

Taj's weekly salary $= 450 + \frac{30}{100} \times$ weekly sales

Ron's weekly salary $= 600 + \frac{15}{100} \times$ weekly sales

s = weekly sales and Taj's weekly salary = Ron's weekly salary

Since they have the same weekly salary, we can set the equations equal to each other.

Step 1:

$$450 + \left(\frac{30}{100}s\right) = 600 + \left(\frac{15}{100}s\right)$$

$$450 + \frac{30s}{100} - 450 = 600 + \frac{15s}{100} - 450$$

$$\frac{30s}{100} = 150 + \frac{15s}{100}$$

$$\frac{30s}{100} - \frac{15s}{100} = 150 + \frac{15s}{100} - \frac{15s}{100}$$

$$\frac{30s}{100} - \frac{15s}{100} = 150$$

$$\frac{15s}{100} = 150$$

$$\frac{15s}{100} \times \frac{150}{1}$$

$$15s = 15000$$

$$\frac{15s}{15} = \frac{15000}{15}$$

$$s = 1000$$

92

Step 2:

Plug in $s = 1000$ to one of the equations:

$$450 + \left(\frac{30}{100} \bullet 1000\right) =$$

$$450 + \frac{30000}{100} =$$

$$450 + 300 = 750$$

Answer: (C) 750

We can also solve by plugging in the answer choices. Plug in (C) 750:

$$750 = 450 + \frac{30}{100}s$$

$$750 - 450 = 450 + \frac{30s}{100} - 450$$

$$300 = \frac{30s}{100}$$

$$\frac{300}{1} \diagdown \frac{30s}{100}$$

$$30s = 30000$$

$$\frac{30s}{30} = \frac{30000}{30}$$

$$s = 1000$$

$$600 + \left(\frac{15}{100} \bullet 1000\right) = 750 \checkmark$$

Both salaries are equal to $750, so (C) works!

Another big part of percents are problems that ask for the ***percent increase*** or ***percent decrease***.

Look for these Key Words: *Percent Increase*, *Percent Decrease*, *Percent More*, *Percent Less*, *Percent Greater*, *Percent Discounted*.

Whenever you see these words use the **Percent Increase/Decrease Formula**: $\dfrac{\text{Change}}{\text{Original}} \times 100$

The *Change* is the difference between the two given numbers.

To determine the Original Number:

Percent Increase: Original Number is the SMALLER NUMBER. (Because you have to have somewhere to increase to.)

Percent Decrease: Original Number is the BIGGER NUMBER. (Because you have to have somewhere to decrease from.)

Let's see how this works:

5. A $60 sweater is on sale for $36. The sale price is what percent less than the normal price?

First find the change: $60 - 36 = 24$

It wants the percent decrease so the original number must be the bigger number: 60

Plug these values into the formula: $\frac{24}{60} \times 100 = \frac{2}{5} \times 100$ or 40%

Let's try a more complicated problem:

11. Bill's salary was decreased by 25 percent. After making several more errors at work, his new salary was decreased by 10 percent. What percent would his salary have to increase to get his salary back to the original amount?

> Explanation:
> This is a Sneaky Plug In problem. Let's throw in 100 for Bill's salary.
> 25 percent of 100 is 25
> $$\frac{25}{100} \cdot 100 = \frac{25}{\cancel{100}} \cdot \frac{\cancel{100}}{1} = 25$$
> Now, reduce 100 by 25, and Bill's new salary is 75.
> 10 percent of 75 is 7.50
> $$\frac{10}{100} \cdot 75 = \frac{1\cancel{0}}{10\cancel{0}} \cdot \frac{75}{1} = \frac{75}{10} = 7.5$$
> So, subtract 7.50 from 75 to get a new salary of 67.50. Now we have the magic words "percent increase." Find the change: $100 - 67.50 = 32.50$
> We are looking for the percent increase, so the original number has to be the smaller number: 67.50. ETS uses the word "original" to throw you off. They want you to put 100 (or whatever number you've plugged in) in the denominator and get the problem wrong! Don't fall for it! Remember the rules: percent increase means the original number (the number in the denominator) must be the smaller number!
>
> Plug these values into the formula: $\frac{32.50}{67.50} \cdot 100 =$ approximately 48%.
>
> Answer: 48

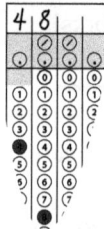

RATIOS

Ratios are simply fractions that express a relationship between two items.

The ratio of 2 to 3 can be written as: 2:3 or $\frac{2}{3}$

I'm sure in school you learned to set up proportions to solve for ratios, however, there is a much easier way to solve for ratios. Just use a Ratio Grid!

Let's see how it works with the following problem:

The ratio of boys to girls in Ms. Chin's math class is 3 to 5. If there are 40 total students in the class, how many girls are in Ms. Chin's math class?

	BOYS	GIRLS	TOTAL
RATIO	3	(+) 5 (=)	8
MULTIPLIER	× 5	× 5	× 5
ACTUAL #	15	25	40

Add the ratio of boys to girls to get the ratio total:
$3 + 5 = 8$

Ask: How do I get from 8 to 40? Multiply by 5.

So there are 25 girls in Ms. Chin's class.

Let's try some more Ratio problems:

10. A total of 143,000 people live in Jefferson County. If the ratio of women to men is 6 to 5, what is the total number of men?

(A) 50,000
(B) 65,000
(C) 78,000
(D) 85,000
(E) 100,000

Explanation:
Use a Ratio Grid:

	Women	Men	Total
Ratio	6 (+)	5 (=)	11
Multiplier		× 13,000	× 13,000
Actual #		65,000	143,000

Ask: How do I get from 11 to 143,000?
or
What is 143,000 divided by 11?

Answer: (B) 65,000

Here's a ratio problem that's a little bit different:

13. A set of toy building blocks comes with 14 green blocks and 14 yellow blocks. What is the least number of blocks that could be removed so that the ratio of green blocks to yellow blocks is 5 to 7?

(A) 2
(B) 3
(C) 4
(D) 5
(E) 6

Explanation:

$$\frac{green}{yellow} = \frac{5}{7}$$

To find the least number of blocks that can be removed, keep the number of yellow blocks as 14 and set up a proportion and cross-multiply:

$$\frac{green}{14} \diagdown \frac{5}{7}$$

$7(green) = 70$

$$\frac{7(green)}{7} = \frac{70}{7}$$

$green = 10$

We started with 14 green and ended up with 10 green, so what's the difference?
$14 - 10 = 4$

Answer: (C) 4

Instead, we could also use a Ratio Grid:

	Green	Yellow	Total
Ratio	5	7	12
Multiplier	× 2	× 2	× 2
Actual #	10	14	24

Ask: How do I get from 7 to 14?

PROBABILITY

Probability problems on the SAT are really quite straightforward. *Probability* means the chance something will happen. It can be expressed by the following fraction:

$$\frac{\text{\# of desired outcomes}}{\text{\# of total outcomes}}$$

The probability of an event happening is always a fraction between 0 and 1. A probability of 1 means the event will always happen and a probability of 0 means it will never happen.

Remember: the probability of the different possible events (or outcomes) have to add up to 1! So, if there are only blue and red marbles and there is a $\frac{5}{7}$ probability of pulling a red one, then the probability of pulling a blue must be $\frac{7}{7} - \frac{5}{7}$, or $\frac{2}{7}$.

Let's see how this works on an SAT problem:

2. A necklace is made up of 30 green beads, 40 brown beads, and 50 black beads. If the necklace is made up of no other beads, and 1 bead is to be removed from the necklace at random, what is the probability that the removed bead will be green?

(A) $\frac{1}{8}$

(B) $\frac{1}{6}$

(C) $\frac{1}{4}$

(D) $\frac{1}{3}$

(E) $\frac{1}{2}$

Explanation:
Probability means desired over total, so put 30 (the number of green beads) over the total number of beads:

Total number of beads = 30 + 40 + 50 = 120

$$\frac{30 \text{ green beads}}{120 \text{ total beads}} = \frac{1}{4}$$

Answer: (C) $\frac{1}{4}$

Many times it is best to write out your combos just to be on the safe side. Let's try this on the following probability problem:

16. Set P contains the elements {3, 4, 5} and Set M contains the elements {8, 9, 10}. If a member is chosen at random from each of the two sets, what is the probability that the product of a member of Set P and a member of Set M is divisible by 4?

(A) $\frac{5}{9}$

(B) $\frac{4}{6}$

(C) $\frac{4}{9}$

(D) $\frac{4}{5}$

(E) $\frac{6}{9}$

Explanation:
Write out all of the possible outcomes.

$3 \cdot 8 = 24$	Yes
$3 \cdot 9 = 27$	No
$3 \cdot 10 = 30$	No
$4 \cdot 8 = 32$	Yes
$4 \cdot 9 = 36$	Yes
$4 \cdot 10 = 40$	Yes
$5 \cdot 8 = 40$	Yes
$5 \cdot 9 = 45$	No
$5 \cdot 10 = 50$	No

5 yeses out of a total of $9 = \frac{5}{9}$

Answer: (A) $\frac{5}{9}$

COMBINATIONS/PERMUTATIONS

Combination/Permutation problems show up on the SAT. They tend to be tricky, so keep in mind, if a problem makes your brain start to throb, skip it! Conquering a combination problem comes down to asking a particular series of questions to arrive at the right answer. Let's start with the basics. Often the best way to tackle a combination problem is to simply write out your possible combinations. Let's try a standard combination problem that is best solved by writing out the possibilities.

15. The integer 5,421 consists of four different integers that all decrease from left to right. How many integers between 6000 and 7000 have digits that are all different and decrease from left to right?

(A) 16
(B) 20
(C) 26
(D) 32
(E) 36

Explanation:
This is an example of a combination problem that is best done by writing out all possible combinations. Be meticulous and organized so you don't miss any.

6543	6532	6521	6510
6542	6531	6520	
6541	6530		
6540			

6432	6421	6410
6431	6420	
6430		

6321	6310
6320	

6210

So, there are 20 possible combinations.

Answer: (B) 20

But sometimes it's best to solve combination problems mathematically. Let's break down the following flow chart.

Combination Flow Chart

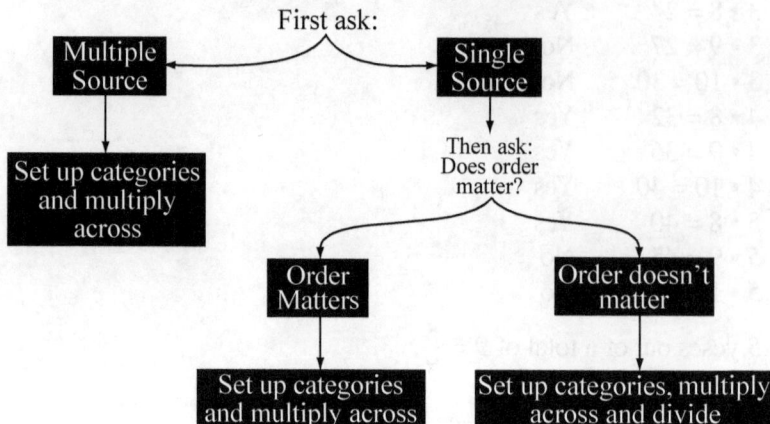

First ask:

```
        Multiple  ←————————→  Single
         Source                Source
            |                    |
            ↓               Then ask:
    Set up categories       Does order
     and multiply            matter?
       across           ┌───────┴───────┐
                        ↓               ↓
                    Order           Order doesn't
                    Matters           matter
                        |               |
                        ↓               ↓
                Set up categories   Set up categories, multiply
                and multiply across   across and divide
```

The first question to ask is: "Is this a Multiple Source Problem or a Single Source Problem."

An example of a ***Multiple Source*** problem is one that involves picking from different sources. For example, "Mark wants to purchase 1 comedy, 1 action, and 1 drama DVD from the video store. He can choose from 20 comedies, 15 action flicks, and 30 dramas." Because there are the different categories of DVDs – comedy, action, and drama – this is a multiple source problem. Had it been worded, "Mark wants to purchase 3 DVDs from the video store. He can choose from 65 DVDs." Then it would be a ***Single Source*** problem because it is only one category of items – DVDs.

If you have determined that you are on a multiple source problem it is simple and easy. Set up your categories and multiply across.

Comedy		Action		Drama		
20	×	15	×	30	=	9000

If you have determined that you are on a ***single source*** problem first set up your categories:

DVD1		DVD2		DVD3	
65	×	64	×	63	=

We have 64 DVDs to choose from for DVD2, because we've already picked one for DVD1, and 63 DVDs available for DVD3, because we've eliminated 2. Now we have to figure out if order matters or if order doesn't matter.

On this problem, let's say Mark chooses *Raiders of the Lost Ark* as his first DVD, and then chooses *Star Wars* second, and *Lord of the Rings* third. If he flip flops the order and chooses *Star Wars* first, then *Lord of the Rings*, and lastly *Raiders of the Lost Ark*, is he picking the same combination of movies, or a different combination of movies? The same one! ***Order doesn't matter***, so we have to divide.

DVD1		DVD2		DVD3		
$\frac{65}{3}$	×	$\frac{64}{2}$	×	$\frac{63}{1}$	=	43,680

You simply divide according to how many positions contain an overlap (it will usually be all or none). 3 DVDs means count down from 3 to 1. If he were picking two DVDs you would count down from 2.

It would look like this: $\frac{65}{2} \times \frac{64}{1}$

For 4 DVDs you would count down from 4, like this: $\dfrac{65}{4} \times \dfrac{64}{3} \times \dfrac{63}{2} \times \dfrac{62}{1}$

Essentially, we are canceling out all those duplicate options. If **order mattered** the question would read more like this: "Mark wants to purchase 3 DVDs from a video store. He can choose from 65 DVDs and is deciding the order in which to watch them."

Now if Mark chooses *Raiders of the Lost Ark*, *Star Wars*, and *Lord of the Rings* as his first combo and then flip flops the order to *Star Wars*, *Lord of the Rings*, and *Raiders of the Lost Ark*, we have a new combo because we have changed the order in which he watches them. When **order matters** we don't divide, so the solution would look like this:

DVD 1		DVD 2		DVD 3		
65	×	64	×	63	=	262,080

If you are comfortable working with formulas and permutations, have at it!

Permutation:

$$nPr = \dfrac{n!}{(n-r)!}$$

Combination:

$$nCr = \dfrac{n!}{(n-r)!\, r!}$$

n stands for the number you are picking from (65 DVDs) and r represents how many items you are choosing (3 DVDs).

Let's try some SAT combination problems:

18. Debbie has a green scarf, a beige scarf, and a brown scarf, as well as 3 belts – one green, one beige, and one brown – and three blouses – one green, one beige, and one brown. Debbie wants her outfit to be green, beige and brown and be made up of one scarf, one belt, and one blouse. How many different possible outfits could she wear?

(A) 27
(B) 12
(C) 9
(D) 6
(E) 3

Explanation:
Recognize that it is a multiple source problem – scarves, belts, and blouses, not simply clothes. Let's first separate out our categories. We have 3 scarves to choose from. Let's say we picked a green one – that leaves only 2 belts to choose from: beige and brown. Let's say we chose brown, and we are left with only one color blouse to choose from, beige.

Scarf		Belt		Blouse		
3	×	2	×	1	=	6

Answer: (D) 6

Alternatively, you can write out all of the possible combinations:

	Scarf	**Belt**	**Blouse**
1.	Green	Beige	Brown
2.	Green	Brown	Beige
3.	Beige	Green	Brown
4.	Beige	Brown	Green
5.	Brown	Green	Beige
6.	Brown	Beige	Green

There are six possibilities.

17. How many different integers made up of four nonzero digits can be formed if the tens digit must be 4, the ones digit cannot be 9, and digits may not be repeated?

(A) 84
(B) 294
(C) 448
(D) 648
(E) 2016

Explanation:

Since there are four digits, draw four blanks: ___ ___ ___ ___

Since the tens digit must be 4, there is only one choice for that blank: ___ ___ _/_ ___

If the ones digit can't be 9, 4 or 0, there are seven options for the ones digit.

___ ___ _/_ _7_

The only restrictions that remain are that the digits cannot be 0 or repeat, so if two digits are

down and the hundreds digit can't be 0 or repeat, there are seven options. ___ _7_ _/_ _7_

For the thousands, there are six options: _6_ _7_ _/_ _7_

multiply the possibilities: $6 \times 7 \times 1 \times 7 = 294$

Answer: (B) 294

PATTERN AND SEQUENCE PROBLEMS

Pattern Problems are *Remainder* problems in disguise. Let's say the problem reads, "Ronnie is beading a necklace in the pattern green, red, blue, black, purple. What color will the 83rd bead be?"

Determine how many terms are in the pattern: green, red, blue, black, purple = 5 terms in the pattern.

So divide 83 by 5:

$$\begin{array}{r} 16 \text{ R3} \\ 5\overline{)83} \\ -5 \\ \hline 33 \\ -30 \\ \hline 3 \end{array}$$

We are left with a remainder of 3, so we are 3 colors into the pattern. The 83rd bead would be blue.
What if we had gotten a remainder of 0? Where would we be in the pattern? At the very end! Purple.

Take a look at the following pattern problem:

12. A baker is baking a batch of cookies using the pattern: tree, star, heart, and dreidel. If the baker shapes his final cookie into a heart, which of the following could be the total number of cookies baked?

(A) 71
(B) 73
(C) 76
(D) 78
(E) 81

Explanation:
Use the A.C.T.! Plug in the answer choices and remember that pattern problems are remainder problems in disguise.

(C) 76 can be eliminated because 76 is divisible by 4 ($76 \div 4 = 19$). We would be at the last shape in the pattern: dreidel.

(B) 73 ($73 \div 4 = 18$ R1) Thus, the last cookie will be the first shape in the pattern: tree.

(A) 71 ($71 \div 4 = 17$ R3) Thus, the last cookie will be the third shape in the pattern: heart.

Answer: (A) 71

Sequence Problems are another variation of Pattern Problems.

The trick with sequence problems is to carry out the sequence until you discover a pattern. From there, pay close attention to what the particular question is asking.

$$-3, 3, 1 \ldots$$

The first term in the sequence of numbers shown above is -3. Each odd numbered term after the first is 2 less than the previous term and each even numbered term is -1 times the previous term. What is the 50th term of the sequence?

Carry out the sequence to determine a pattern:

$-3,$	$3,$	$1,$	$-1,$	$-3,$	$3 \ldots$
term 1	term 2	term 3	term 4	term 5	term 6

-3 is the first term, the 2nd term is an even numbered term so multiply by -1, the 3rd term is an odd numbered term so subtract 2, and the 4th term is an even numbered term so multiply by -1, and the 5th term is an odd numbered term so subtract 2…

We can see that the pattern repeats every four terms.

Pattern Problems are remainder problems in disguise so divide 50 by 4 to produce the remainder of 2:

$$\begin{array}{r} 12\,\text{R2} \\ 4\overline{)50} \\ \underline{-4} \\ 10 \\ \underline{-8} \\ 2 \end{array}$$
We are two terms into the pattern so the 50th term is 3.

Try a sequence problem on your own:

12. Each term of a sequence is less than the term before it. The difference between any two consecutive terms in the sequence is always the same number. If the fourth and eighth terms of the sequence are 123 and 91, respectively, what is the tenth term?

Explanation:

123 〰〰〰 91

first second third fourth fifth sixth seventh eighth ninth tenth

$123 - 91 = 32$

There are four jumps between the 4th and 8th terms.

$$\frac{32}{4} = 8$$

$91 - (2 \bullet 8) =$
$91 - 16 = 75$

Answer: 75

PROPORTIONS

There are two types of proportions tested on the SAT.

Direct Proportion and ***Indirect/Inverse Proportion***

Let's talk about the most common proportion first:

Direct Proportion $\dfrac{x_1}{y_1} = \dfrac{x_2}{y_2}$

When x increases, y increases at the same rate.

Let's try a few:

7. The amount of ice that an ice maker can produce varies directly with the amount of time for which the machine runs. If the ice maker makes 30 pounds of ice in 45 minutes, then how many pounds of ice can the machine make in one hour?

> Explanation:
> Notice the expression "varies directly." That always means set up a proportion. Keep the units of the proportion consistent. In this case we would have:
>
> $$\frac{\text{lbs of ice}}{\text{minutes}} = \frac{\text{lbs of ice}}{\text{minutes}}$$
>
> $$\frac{30}{45} = \frac{x}{60}$$
>
> Be sure to convert one hour into 60 minutes in order to keep the units consistent!
>
> $\dfrac{30}{45} \diagdown\!\!\!\!\diagup \dfrac{x}{60}$ cross-multiply
>
> $$45x = 1800$$
>
> $$\frac{45x}{45} = \frac{1800}{45}$$
>
> $x = 40$ lbs of ice in 1 hour.
>
> Answer: 40

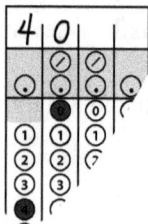

Here's another proportion problem:

8. It takes Louis 2 hours to ride a distance of 18 miles on his bike. If Tom rides his bike at twice this rate, how many minutes would it take Tom to ride a distance of 15 miles?

Explanation:

They tell us Louis can ride 18 miles in 2 hours. Let's figure out how far he can go in one hour:

$$\frac{18 \text{ miles}}{2 \text{ hours}} = \frac{9 \text{ miles}}{1 \text{ hour}}$$

Tom rides at twice this rate:

$9 \times 2 = 18$ miles in 1 hour.

The question asks for minutes, not hours, so we need to convert to 18 miles in 60 minutes.

$$\frac{18 \text{ miles}}{1 \text{ hour}} \bullet \frac{1 \text{ hour}}{60 \text{ minutes}} = \frac{18 \text{ miles}}{60 \text{ minutes}}$$

Now, write a proportion:

$$\frac{18 \text{ miles}}{60 \text{ minutes}} \diagdown\diagup \frac{15 \text{ miles}}{x \text{ minutes}}$$

$$18x = 900$$

$$\frac{18x}{18} = \frac{900}{18}$$

$$x = 50 \text{ minutes}$$

Answer: 50

The other type of proportion problem that appears on the SAT is the ***Inverse*** or ***Indirect Proportion*** problem.

Inverse/Indirect Proportion $x_1 y_1 = x_2 y_2$

In this type of situation, when x increases, y decreases and vice versa.

Let's say it takes 5 waiters 2 hours to clean a restaurant. At this rate, how long does it take 10 waiters to clean a restaurant?

Just think about it common sense wise – If you double the amount of waiters on cleaning duty, then you lessen the amount of time it takes to clean the restaurant. Plug the given numbers into the formula to see how it works.

$$5(2) = 10x$$

$$10 = 10x$$

$$\frac{10}{10} = \frac{10x}{10}$$

$$x = 1 \text{ hour}$$

Let's try one that uses the words inverse or indirect proportion.

If x varies inversely with y and x is 3 when y is 12, what is x when y is 9?

$$3(12) = 9x$$

$$36 = 9x$$

$$\frac{36}{9} = \frac{9x}{9}$$

$$x = 4$$

CHARTS AND TABLES

The only other thing to be aware of with these basic arithmetic and algebra topics is that often they are tested in the form of a chart or a table.

Take the time to study the chart or table given before you proceed to do any math. Often, the chart seems easy and simple enough, but beware, there is always something tricky lurking under the surface.

Let's try a couple:

Year	Employees
2000	1,700
2001	1,875
2002	2,015
2003	2,190
2004	2,330

16. The number of employees working at company A from 2000 through 2004 is shown in the chart above. Beginning in 2000, the number of employees has increased by a constant every 2 years. How many employees worked at company A in the year 2006?

(A) 2470
(B) 2505
(C) 2645
(D) 2785
(E) 2820

Explanation:
Work out the difference between each year to discover the pattern.

2000 to 2001: 1,875 – 1,700 = 175
2001 to 2002: 2,015 – 1,875 = 140
2002 to 2003: 2,190 – 2,015 = 175
2003 to 2004: 2,330 – 2,190 = 140

Between 2000 and 2001 the number of employees increases by 175. From 2001 to 2002 it increases by 140. The chart lists the number of employees per year so the constant every two years is 175 + 140 = 315. That's the trick! Many students will subtract the difference of one year, not two! Add 315 to 2330 (315 + 2330 = 2645).

Answer: (C) 2645

City	Number of Consecutive Days
A	7
B	3
C	5
D	8
E	6
F	7

16. The table above shows the number of consecutive days that each of five cities in Rocksford County lost power during a twelve-day period. If City C's power outage did not overlap with City F's power outage, which of the 12 days could be a day when only one city lost power?

(A) The 4th
(B) The 5th
(C) The 6th
(D) The 7th
(E) The 8th

Explanation:

Days:	1st	2nd	3rd	4th	5th	6th	7th	8th	9th	10th	11th	12th
						A	A	A	A	A	A	A
										B	B	B
							C	C	C	C	C
				D	D	D	D	D	D	D	D
						E	E	E	E	E	E
	F	F	F	F	F	F	F					

Answer: (A) The 4th

Let's move on to the Algebra lesson.

Chapter 8
Algebra Lesson

It's important to be comfortable manipulating *algebraic equations*.

Let's try several to test your knowledge:

1. $5y = y + 12$

$$5y - y = y + 12 - y$$
$$4y = 12$$
$$\frac{4y}{4} = \frac{12}{4}$$
$$y = 3$$

2. $\frac{7x}{3} = 14$

$$\frac{7x}{3} = \frac{14}{1}$$

Cross-multiply

$$\frac{7x}{3} \diagdown \frac{14}{1}$$
$$7x = 42$$
$$\frac{7x}{7} = \frac{42}{7}$$
$$x = 6$$

3. $\frac{1}{4x - 6} = \frac{3}{26 + x}$

Cross-multiply

$$\frac{1}{4x-6} \diagup\!\!\!\!\diagdown \frac{3}{26+x}$$

$$3(4x - 6) = 26 + x$$
$$12x - 18 = 26 + x$$
$$12x - 18 - x = 26 + x - x$$
$$11x - 18 = 26$$
$$11x - 18 + 18 = 26 + 18$$
$$11x = 44$$
$$\frac{11x}{11} = \frac{44}{11}$$
$$x = 4$$

4. $\dfrac{\frac{x}{3}}{\frac{x}{15}} = 13 - x$

$$\frac{x}{3} \div \frac{x}{15} = 13 - x$$

Keep it   Switch it   Flip it

$$\frac{x}{3} \times \frac{15}{x} = 13 - x$$
$$\frac{15x}{3x} = 13 - x$$
$$\frac{15x}{3x} = 13 - x$$
$$5 = 13 - x$$
$$5 - 13 = 13 - x - 13$$
$$-8 = -x$$

Divide by −1 to make x positive:

$$\frac{-8}{-1} = \frac{-x}{-1}$$
$$x = 8$$

5. $\frac{4x^6 + 4x^8}{2x^6} = 34$

$$\frac{4x^6 + 4x^8}{2x^6} = \frac{4x^6}{2x^6} + \frac{4x^8}{2x^6} = 34$$

Divide coefficients and subtract exponents:

$$2x^0 + 2x^2 = 34$$

Any base raised to the power of 0 is 1.

$$2 \bullet 1 + 2x^2 = 34$$
$$2 + 2x^2 = 34$$
$$2 + 2x^2 - 2 = 34 - 2$$
$$2x^2 = 32$$
$$\frac{2x^2}{2} = \frac{32}{2}$$
$$x^2 = 16$$
$$\sqrt{x^2} = \sqrt{16}$$
$$x = \pm 4$$

Let's try an SAT problem:

11. If $\dfrac{a+5b}{a} = \dfrac{3}{2}$, what is the value of $\dfrac{a}{b}$?

(A) 15
(B) 10
(C) 12
(D) 2
(E) 22

Explanation:

$$\dfrac{a+5b}{a} = \dfrac{3}{2}$$

Cross-multiply:

$2(a + 5b) = 3a$

$2a + 10b = 3a$

$2a + 10b - 2a = 3a - 2a$

$10b = a$

$$\dfrac{10b}{b} = \dfrac{a}{b}$$

$$10 = \dfrac{a}{b}$$

Answer: (B) 10

On the SAT, many algebraic equation problems require you to translate English into Math. For instance, if ETS says, "a number times itself is two more than six times y," you would need to be able to write that as a mathematical equation. Let's try it:

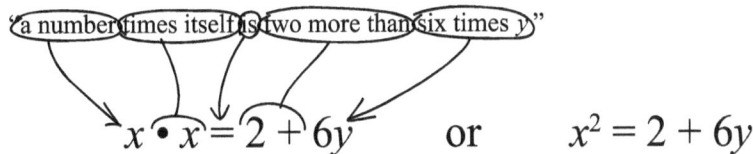

"a number times itself is two more than six times y"

$$x \bullet x = 2 + 6y \qquad \text{or} \qquad x^2 = 2 + 6y$$

How about these:

If twice a number subtracted from 10 is 16, what is the number?

$$10 - 2n = 16$$
$$10 - 2n - 10 = 16 - 10$$
$$-2n = 6$$
$$\dfrac{-2n}{-2} = \dfrac{6}{-2}$$
$$n = -3$$

Eight less than four times a number is 92. What is the number?

$$4n - 8 = 92$$
$$4n - 8 + 8 = 92 + 8$$
$$4n = 100$$
$$\dfrac{4n}{4} = \dfrac{100}{4}$$
$$n = 25$$

5 times the sum of 40 and a number is 250. What is the number?

$$5(40+x) = 250$$
$$200 + 5x = 250$$
$$200 + 5x - 200 = 250 - 200$$
$$5x = 50$$
$$\frac{5x}{5} = \frac{50}{5}$$
$$x = 10$$

Let's try some SAT problems:

6. Which of the following expressions represents the statement, "y is 3 more than the product of x and y"?

(A) $y = 3 + (x + y)$
(B) $y = 3 + xy$
(C) $3xy = y$
(D) $3 + xy > y$
(E) $xy + y > 3$

Explanation:
Just translate the words into math symbols: 'y' is y, 'is' stands for $=$, '3 more' means $+$, so $3 +$, the 'product' means the result of a multiplication problem, and so forth.

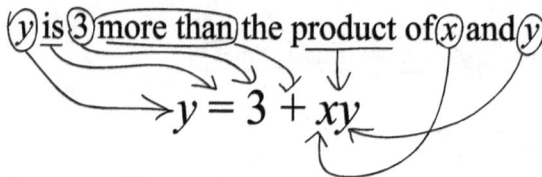

y is 3 more than the product of x and y

$$y = 3 + xy$$

Answer: (B) $y = 3 + xy$

---

8. $a, b,$ and c represent three numbers whose average (arithmetic mean) is 14. When the sum of the two smallest numbers is subtracted from the greatest number, the result is 8. If $a > b > c$, which of the following pairs of equations could correctly express the information above?

(A) $\begin{array}{l} a+b+c = 42 \\ a-b-c = 8 \end{array}$

(B) $\begin{array}{l} a+b+c = 42 \\ c-a+b = 8 \end{array}$

(C) $\begin{array}{l} a+b+c = 42 \\ c-a+b = 16 \end{array}$

(D) $\begin{array}{l} a+b+c = 28 \\ c-a+b = 8 \end{array}$

(E) $\begin{array}{l} a+b+c = 28 \\ a-b+c = 16 \end{array}$

Explanation:
Step 1:
Plug $a, b,$ and c into the formula for average to find the sum:

$$\frac{a+b+c}{3} = 14$$

$$\frac{a+b+c}{3} = \frac{14}{1}$$

$$a+b+c = 42$$

$$42 \neq 28$$

So, eliminate (D) and (E)

Step 2:
$a > b > c$, so
b and c are the smallest numbers, and a is the largest. So,
$$a - (b + c) = 8$$
distribute the minus sign to b and c:
$$a - b - c = 8$$

Answer: (A) $\begin{array}{l} a+b+c = 42 \\ a-b-c = 8 \end{array}$

SIMULTANEOUS EQUATIONS

You probably learned to deal with *Simultaneous Equations* using substitution. On the SAT, elimination is better because they usually give you really easy numbers. First you must be able to recognize a simultaneous equation problem.

How to Identify: Two or More Equations with Two or More Variables.

How to Solve: Put One Equation on Top of the other and either Add or Subtract.

Let's say you are working with the following equations:

$$5x + 4y = 9 \text{ and } 3x - 4y = 7$$

First, stack them on top of each other and add to cancel out the y and solve for the x:

$$
\begin{array}{r}
5x + 4y = 9 \\
+ \quad 3x - 4y = 7 \\
\hline
8x + \ 0 = 16 \\
\dfrac{8x}{8} = \dfrac{16}{8} \\
x = 2
\end{array}
$$

Now, plug in the x value to determine y:

$$
\begin{aligned}
5x + 4y &= 9 \\
5(2) + 4y &= 9 \\
10 + 4y &= 9 \\
10 + 4y - 10 &= 9 - 10 \\
4y &= -1 \\
\frac{4y}{4} &= \frac{-1}{4} \\
y &= -\frac{1}{4}
\end{aligned}
$$

Try this pair of equations: $3x - 9y = -18$ and $x + y = 10$

<u>Step 1:</u>

$$
\begin{array}{r}
x + y = \ 10 \\
+ \ 3x - 9y = -18 \\
\hline
4x - 8y = -8
\end{array}
$$

<u>Step 2:</u>

Neither variable cancels, so multiply the first equation by 9 before adding:

$$9(x + y) = 9(10)$$

$$9x + 9y = 90$$

<u>Step 3:</u>

Now stack and add the variables to eliminate the y:

$$
\begin{array}{r}
3x - 9y = -18 \\
+ \quad 9x + 9y = 90 \\
\hline
12x + \ 0 = 72 \\
\dfrac{12x}{12} = \dfrac{72}{12} \\
x = 6
\end{array}
$$

You might have to multiply out by a negative after adding the equations.

Try this one: If $x - y = 5$, and $2x + z = 10$, what is the value of $x + y + z$?

Explanation:

When subtracting simultaneous equations be sure to distribute the minus sign to every term in the equation.

$$
\begin{array}{l}
\ x - y + 0 = 5 \\
-\ \ 2x + 0 + z = 10 \\
\hline
\ -x - y - z = -5 \\
-1(-x - y - z) = -1(-5) \\
x + y + z = 5
\end{array}
$$
Put in 0 as a place holder in order to line up the variables.

Look what happens if we add instead of subtract:

$$
\begin{array}{l}
\ x - y + 0 = 5 \\
+\ \ 2x + 0 + z = 10 \\
\hline
\ 3x - y + z = 15
\end{array}
$$

Doesn't do us much good!

Let's try some SAT problems:

$$4c + 5d + 4f = 21$$
$$c + 5d + f = 15$$

7. If the above equations are true, which of the following is a value of $c + f$?

(A) –2
(B) 0
(C) 2
(D) 3
(E) 6

Explanation:

Don't forget to distribute the minus sign to each term.

$$
\begin{array}{l}
\ 4c + 5d + 4f = 21 \\
-\ \ (c + 5d + f) = 15 \\
\hline
\ 3c + 0 + 3f = 6
\end{array}
$$
$$\frac{3c + 3f}{3} = \frac{6}{3}$$
$$c + f = 2$$

Answer: (C) 2

Let's try a word problem that involves Simultaneous Equations:

11. It costs $12.00 for an adult ticket and $2.00 for a child ticket to the Renaissance Fair. If 400 tickets were sold for a total of $2400, what is the ratio of child to adult tickets sold?

(A) $\frac{1}{12}$

(B) $\frac{1}{6}$

(C) $\frac{1}{2}$

(D) $\frac{3}{2}$

(E) $\frac{3}{4}$

Explanation:

Let a equal the number of adult tickets sold and c equal the number of child tickets sold.

$$12a + 2c = 2400 \qquad\qquad a + c = 400$$

Translation:
$12 for each adult ticket plus $2 for each child ticket equals $2,400 total for tickets sold.

Translation:
The number of adult tickets plus the number of child tickets equals 400 total tickets sold.

Notice that if we add the equations as is, none of the variables will cancel, so multiply the second by -2 before adding: $-2(a+c) = -2(400)$

$$-2a + -2c = -800$$

Add the two equations together:

$$\begin{array}{r} 12a + 2c = 2400 \\ + \quad -2a + -2c = -800 \\ \hline 10a + 0 = 1600 \end{array}$$

$$10a = 1600$$
$$\frac{10a}{10} = \frac{1600}{10}$$
$$a = 160$$
$$160 + c = 400$$
$$160 + c - 160 = 400 - 160$$
$$c = 240$$

$$\frac{c}{a} = \frac{240}{160} = \frac{3}{2}$$

Answer: (D) $\dfrac{3}{2}$

ABSOLUTE VALUE

We've already discussed **Absolute Value** in the Math Vocabulary Lesson but let's see how to deal with Absolute Value equations. Most absolute value problems have two parts to the answer.

$$|3x + 5| + 1 = 12$$

Subtract the 1 from both sides in order to get the absolute value expression alone on the left.

$$|3x + 5| + 1 - 1 = 12 - 1$$
$$|3x + 5| = 11$$

Now drop the absolute value sign and rewrite the equation two ways. One way, set the equation equal to 11, and the other way, multiply the right side by (-1) and therefore set the equation equal to -11.

$$3x + 5 = 11 \qquad \text{or} \qquad 3x + 5 = -11$$
$$3x + 5 - 5 = 11 - 5 \qquad\qquad 3x + 5 - 5 = -11 - 5$$
$$3x = 6 \qquad\qquad\qquad 3x = -16$$
$$\frac{3x}{3} = \frac{6}{3} \qquad\qquad\qquad \frac{3x}{3} = \frac{-16}{3}$$
$$x = 2 \qquad\qquad\qquad\qquad x = \frac{-16}{3}$$

Absolute Value Inequalities

$$|2x + 2| < 3$$

$2x + 2 < 3$ or $-(2x + 2) < 3$ which is equivalent to:

$2x + 2 - 2 < 3 - 2$ $-2x - 2 < 3$ $2x + 2 > -3$ ← This way is easier, so take the shortcut and multiply the right side by -1. Remember to flip the sign!

$2x < 1$ $-2x - 2 + 2 < 3 + 2$ $2x + 2 - 2 > -3 - 2$

$x < \dfrac{1}{2}$ $-2x < 5$ $2x > -5$

When dividing by a negative, remember to flip the sign! $\dfrac{-2x}{-2} > \dfrac{5}{-2}$ $\dfrac{2x}{2} > -5$

$x > -\dfrac{5}{2}$ $x > -\dfrac{5}{2}$

Let's try an SAT problem:

14. If $|-3x + 4| < 4$, what is one possible value of x?

Explanation:

$-3x + 4 < 4$ or $-3x + 4 > -4$

$-3x + 4 - 4 < 4 - 4$ $-3x + 4 - 4 > -4 - 4$

Dividing by a negative, so remember to flip the sign! $-3x < 0$ $-3x > -8$

$\dfrac{-3x}{-3} > 0$ $\dfrac{-3x}{-3} < \dfrac{-8}{-3}$ (Flip the sign!)

$x > 0$ $x < \dfrac{8}{3}$

Answer: $0 < x < \dfrac{8}{3}$

This is a grid in, so grid in any value of x between 0 and $\dfrac{8}{3}$, such as 2.

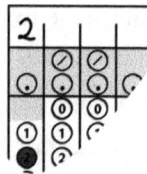

You should know that if ETS asks what the graph of an absolute value looks like, it is a V shape. Like this:

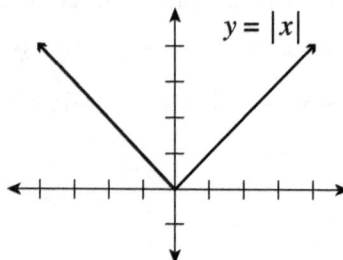

$y = |x|$

EXPONENTS

Exponents and *Roots* are all over the test so let's review our rules.

First we need to know some basic terminology.

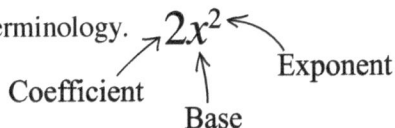

$$2x^2$$

Coefficient · Base · Exponent

What about 3^2? Is 3 a coefficient now? No, it is the base! The base is the term being raised to an exponent.

Adding and Subtracting with Exponents

The Rule: Must have the same base and exponent.
To Solve: Add the coefficients. Base and exponent stay the same.

Let's try some:
$$2x^2 + 3x^2 = 5x^2$$
$$6x^3 - 2x^3 = 4x^3$$
$$5x^3 - 3x^2 = \text{CAN'T DO} \text{ (They don't have the same exponent!)}$$

Let's test this concept in a more difficult fashion:
$$y^2 + y^2 + y^2 = 3y^2$$

There are invisible 1s hanging out in front of the y so we add these 1s because they are coefficients and the base and exponent (y^2) stay the same.

$$(1)y^2 + (1)y^2 + (1)y^2 = 3y^2$$

Let's try the same thing but with a number as the base and a variable as the exponent:
$$3^x + 3^x + 3^x =$$
$$(1)3^x + (1)3^x + (1)3^x = 3^1(3^x)$$
$$3^1 \cdot 3^x = 3^{x+1}$$

In this last instance the 3 is the base and remember the base and exponent stay the same so we know we have to keep 3^x intact. However, there are invisible 1s hanging out in front of the 3s so we add them to give us a 3 in front of the 3^x just as we had a 3 in front of y^2. What is happening between the 3 and the y^2 and the 3 and the 3^x? They are being multiplied. Our rule for multiplication is to check that they have the same base (in this case 3) and then to add the exponents: $x + 1$.

Multiplying with Exponents

The Rule: Must have same base.
To Solve: Multiply the coefficients. Add the exponents. Base stays the same.

Let's try a couple:
$$10x^4 \cdot 5x^2 = 50x^6$$
$$2^3 \cdot 2^5 = 2^8$$

Dividing with Exponents

The Rule: Must have same base.
To Solve: Divide the coefficients and subtract the exponents.

Examples:
$$\frac{24x^4}{6x} = 4x^3$$

$$\frac{7x^7}{21x^3} = \frac{x^4}{3}$$

113

Raising to the Power

To Solve: Multiply exponents. Make sure you also raise the coefficient to the given power.

$$(2x^2)^3 = 8x^6$$

You also need to know what to do when a base is raised to a **negative exponent**. Just put the base and exponent in the denominator, and make the exponent positive and vice versa. Take a look:

$$x^{-2} = \frac{1}{x^2} \qquad 5^{-3} = \frac{1}{5^3} \qquad \frac{1}{2y^{-4}} = \frac{y^4}{2}$$

What if a base is raised to a fraction? The number in the numerator is always the power, and the number in the denominator is always its root.

$$x^{\frac{1}{2}} = \sqrt{x} \qquad x^{\frac{2}{3}} = \sqrt[3]{x^2} \qquad 4^{\frac{3}{4}} = \sqrt[4]{4^3} = \sqrt[4]{64} \qquad x^{-\frac{2}{3}} = \frac{1}{x^{\frac{2}{3}}} = \frac{1}{\sqrt[3]{x^2}}$$

You can always put these in your calculator. Just be sure to use parentheses.

What if a base is raised to the power of 0? Anything raised to the power of 0 is 1.

$$5^0 = 1 \qquad 1{,}349^0 = 1$$

Common Bases

By **common base** I mean that 2, 8, and 16 can all be broken down into a common base of 2; and 5, 25, and 125 all share the common base of 5. You want to be aware of the following:

x^0	x^1	x^2	x^3	x^4	x^5
1	2	4	8	16	32
1	3	9	27	81	
1	4	16	64	256	
1	5	25	125	625	
1	6	36	216		

Let's say ETS gives you the following: $2^{2y} \times 4^3 = 16$ and asks you to solve for y.

First recognize that 2, 4, and 16 all have the common base of 2. Break 4 down into 2^2 and 16 down into 2^4:

$$2^{2y} \times 4^3 = 16$$
$$2^{2y} \times 2^{(2)(3)} = 2^4$$
$$2^{2y} \times 2^6 = 2^4$$
$$2^{2y+6} = 2^4$$

Now each term has the common base of 2 so we can remove the bases and solve the basic algebraic equation created by the exponents.

$$2y + 6 = 4$$
$$2y + 6 - 6 = 4 - 6$$
$$2y = -2$$
$$\frac{2y}{2} = \frac{-2}{2}$$
$$y = -1$$

Plug your y back in to check your answer:

$$2^{2y} \times 4^3 = 16$$
$$2^{2 \cdot (-1)} \times 4^3 = 16$$
$$2^{-2} \times 4^3 = 16$$
$$\tfrac{1}{4} \times 64 = 16$$
$$\frac{64}{4} = 16$$
$$16 = 16 \checkmark$$

Let's try an SAT problem:

12. If $3^x \cdot 3^y = 81$, $(3^x)^y + (3^x)^y + (3^x)^y = 243$, and x and y are positive integers, which of the following could be y?

(A) $\frac{1}{2}$

(B) $\frac{1}{3}$

(C) 1

(D) 2

(E) 3

Explanation:

The first thing to notice is that this is a common base problem. The integers 81 and 243 can all be written in terms of the power of 3. So let's rewrite the equations:

$3^x \cdot 3^y = 81$
$3^x \cdot 3^y = 3^4$

Once the equation is written in terms of the common base, we can drop the base and work with our exponent rules. When multiplying with exponents we add, so $x + y = 4$.

Do the same with the other equation:

$(3^x)^y + (3^x)^y + (3^x)^y = 243$
$3^{xy} + 3^{xy} + 3^{xy} = 3^5$

Rules for addition: if they have the same base, add the coefficients. Remember those invisible 1s! Base and exponent stay the same, so we have $3(3^{xy}) = 3^5$ which is the same as $3^1 \cdot 3^{xy}$, so $1 + xy = 5$.

$$1 + xy = 5$$
$$1 + xy - 1 = 5 - 1$$
$$xy = 4$$

Now let's plug in (C) 1 for y into the two equations:

Equation 1: | Equation 2:
$xy = 4$ | $x + y = 4$
$x(1) = 4$ | $4 + 1 = 5$
$x = 4$ | $5 \neq 4$ ✗

Plug in (D) 2 for y:

Equation 1: | Equation 2:
$xy = 4$ | $x + y = 4$
$x(2) = 4$ | $2 + 2 = 4$
$2x = 4$ | $4 = 4$ ✓
$\dfrac{2x}{2} = \dfrac{4}{2}$ |
$x = 2$ |

Answer: (D) 2

ROOTS

Your calculator is your best friend when it comes to ***square roots***; however, let's go over the basic rules.

Simplifying square roots: we all pretty much know our perfect squares. The square root of 4 is 2, the square root of 36 is 6, but it is important to know how to simplify square roots that are not perfect squares.

Let's break down square root 18

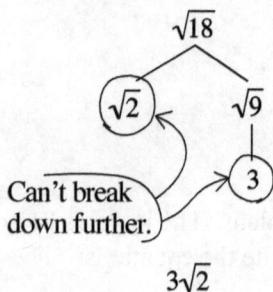

$$\sqrt{18}$$

$\sqrt{2}$ $\sqrt{9}$

Can't break down further.   3

$3\sqrt{2}$

Now try square root 27

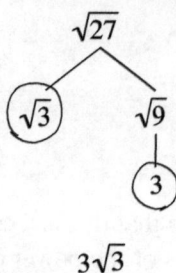

$$\sqrt{27}$$

$\sqrt{3}$ $\sqrt{9}$

3

$3\sqrt{3}$

How about square root 108

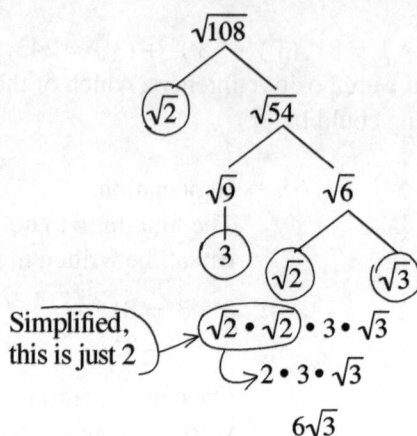

$$\sqrt{108}$$

$\sqrt{2}$ $\sqrt{54}$

$\sqrt{9}$ $\sqrt{6}$

3   $\sqrt{2}$ $\sqrt{3}$

Simplified, this is just 2 $\sqrt{2} \cdot \sqrt{2}) \cdot 3 \cdot \sqrt{3}$

$2 \cdot 3 \cdot \sqrt{3}$

$6\sqrt{3}$

Another trick ETS implements is that they will never leave a number with a square root in the denominator. They will always *rationalize the denominator* by multiplying both the numerator and denominator by the square root given in the denominator.

For instance: $\dfrac{2}{\sqrt{7}} \cdot \dfrac{\sqrt{7}}{\sqrt{7}} = \dfrac{2\sqrt{7}}{7}$

Let's talk about our rules for adding, subtracting, multiplying, and dividing with square roots.

Adding and Subtracting

Rule: Must have same term under the radical.
To Solve: Add or subtract coefficients. The number under the radical stays the same.

For instance: $2\sqrt{3} + 8\sqrt{3} = 10\sqrt{3}$ and $9\sqrt{x} - \sqrt{x} = 8\sqrt{x}$

Multiplying with Square Roots

To Solve: Multiply the numbers outside the radical with each other and multiply the numbers inside the radical with each other.

For instance: $3\sqrt{5} \times 4\sqrt{2} = 12\sqrt{10}$

Dividing with Square Roots

To Solve: Divide the numbers outside the radical with each other and divide the numbers inside the radical with each other.

For instance: $\dfrac{5\sqrt{10}}{35\sqrt{2}} = \dfrac{1\sqrt{5}}{7} = \dfrac{\sqrt{5}}{7}$

QUADRATICS AND FACTORING

Here are the equations you need to know for Quadratics and Factoring:

$$x^2 + 2xy + y^2 = (x+y)(x+y) \text{ or } (x+y)^2$$
$$x^2 - 2xy + y^2 = (x-y)(x-y) \text{ or } (x-y)^2$$
$$x^2 - y^2 = (x+y)(x-y)$$

When given a quadratic in its factored form use FOIL to expand it; if already expanded, factor it. FOIL stands for First, Outer, Inner, Last. Let's try a couple:

$$(x+15)(x-9)$$
$$x^2-9x+15x-135$$
$$x^2+6x-135$$

$$(x-7)(x+7)$$
$$x^2+7x-7x-49$$
$$x^2-49$$

$$(2x+3)(5x+10)$$
$$10x^2+20x+15x+30$$
$$10x^2+35x+30$$

Now let's work with factoring:

1. $x^2+18x-40$

Step 1:
Separate into factors and identify your signs. Whenever a quadratic has a (+) as the first sign, the signs remain the same as in the quadratic:
$$(x+\ )(x-\ )$$

Step 2:
Because you have two different signs $(+,-)$, you know you are looking for a difference. The middle coefficient is 18, so you are looking for a difference of 18.

Factors of 40:
40, 1 : 39
10, 4 : 6
20, 2 : 18
8, 5 : 3

Step 3:
You are looking for a (+) 18 so put the bigger number (in this case 20) by the (+) sign.

$$(x+20)(x-2)$$

2. x^2+5x+6

Step 1:
The first sign is a (+), so the signs remain the same:
$$(x+\ )(x+\ )$$

Step 2:
You have two of the same sign $(+,+)$, so you know you are looking for a sum of 5.

Factors of 6:
6, 1 : 7
2, 3 : 5

Step 3:
$$(x+2)(x+3)$$

3. $x^2-2x-24$

Step 1:
You have a (−) and a (−), which becomes $(+,-)$ inside the parentheses:
$$(x+\ )(x-\ )$$

Step 2:
You have different signs $(+,-)$, so you know you are looking for a difference of 2.

Factors of 24:
24, 1 : 23
6, 4 : 2
8, 3 : 5
12, 2 : 10

Step 3:
You are looking for a (−) 2 so put the bigger number (6) in front of the (−) sign.

$$(x+4)(x-6)$$

If Quadratics aren't your thing don't worry; many of these problems can be solved with plugging in.

Let's try an SAT problem:

12. If $(3x+4)(3x-4)=4$, what is the value of $9x^2$?

(A) −20
(B) −10
(C) 10
(D) 20
(E) 30

Explanation:
Use the FOIL method:

$$(3x+4)(3x-4)=4$$
$$9x^2-12x+12x-16=4$$
$$9x^2-16=4$$
$$9x^2-16+16=4+16$$
$$9x^2=20$$

Answer: (D) 20

There is a sneaky trick ETS likes to employ on Quadratic problems and it involves the expression $x^2 + y^2$. Students will spend valuable time trying to factor this expression. While the expression $x^2 - y^2$ is factorable as $(x + y)(x - y)$, the expression $x^2 + y^2$ is not. When you see this expression what ETS really wants you to do is substitute in a number. Let's see how this works:

If $xy = 45$ and $x^2 + y^2 = 100$, what is the value of $(x - y)^2$?

First, expand out $(x - y)^2$: $x^2 - 2xy + y^2$

Now, regroup the terms: $x^2 + y^2 - 2xy$

Now substitute in 100 for $x^2 + y^2$ and 45 for xy: $100 - 2(45) = 10$.

GROUP PROBLEMS

Group Problems can be solved using a Venn Diagram or the following formula:

$$\text{Total} = \text{Group 1} + \text{Group 2} + \text{Neither} - \text{Both}.$$

Let's see how this works:

There are 200 members of the country club. 80 members play golf, 50 members play tennis, and 20 members play both golf and tennis. How many members play neither golf nor tennis?

Just plug these values into the formula: Total = Group 1 + Group 2 + Neither – Both

$$200 = 80 + 50 + \text{Neither} - 20$$
$$200 = 130 + \text{Neither} - 20$$
$$200 = 110 + \text{Neither}$$
$$200 - 110 = 110 + \text{Neither} - 110$$
$$90 = \text{Neither}.$$

Often ETS will not incorporate a Neither into the question. If this is the case, then still utilize the formula, but without the Neither: Total = Group 1 + Group 2 – Both

Let's do a couple:

9. Sally is throwing herself a birthday party and has prepared 42 gift bags for her guests. Each gift bag contains a candle only, stationary only, or both a candle and stationary. If 27 of the bags contain candles and 24 of the bags contain stationary, how many contain both a candle and stationary?

    Explanation:
    Use the formula:
    Total = Group 1 + Group 2 + Neither – Both.
    Now remove the Neither as it is not a component of the question.

    $$42 = 27 + 24 - \text{Both}$$
    $$42 = 51 - \text{Both}$$
    $$42 - 51 = 51 - \text{Both} - 51$$
    $$-9 = -\text{Both} \quad \text{(Divide both sides by } -1 \text{ to cancel out the negative)}$$
    $$9 = \text{Both}$$

Answer: 9

---

15. Of the members of Sunnyslope Country Club, 83 play golf and 112 play tennis. Some of the members of the country club play both golf and tennis. If 62 members play tennis but not golf, how many members play golf, but not tennis?

Explanation:
Organize your information first: 83 = GOLF, 112 = TENNIS, 62 = ONLY tennis

```
  112 = TENNIS
−  62 = ONLY tennis
   50 = TENNIS and GOLF (Both)
```

```
   83 = GOLF
−  50 = TENNIS and GOLF
   33 = ONLY golf
```

Answer: 33

Let's practice some of these concepts in a quick drill of seven problems.

Algebra Drill

5. If 5 more than twice a number is equal to 20, what is 6 times the number?

(A) 25
(B) 30
(C) 30 ½
(D) 45
(E) 75

9. If $(x - y)^2 = 25$ and $(x + y)^2 = 81$, what is the value of xy ?

(A) 9
(B) 14
(C) 24
(D) 27
(E) 40

10. If $2^{4a} \cdot 2^{2b} = 64$, and a and b are positive integers, what is the value of $2a + b$?

(A) $\frac{3}{2}$
(B) 2
(C) 3
(D) 4
(E) 6

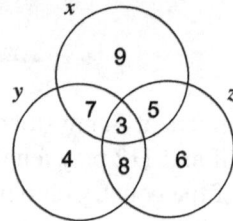

11. The Venn diagram above represents sets x, y, and z where the number in each region indicates the number of elements in that region. How many elements are common to sets y and z ?

(A) 33
(B) 23
(C) 12
(D) 11
(E) 8

15. If $b > 1$ in the equation $b^x \cdot b^{-5} = b^{-y}$ and $b^y \div b^{-3} = b^x$, what is the value of x?

16. Alex removes 7 white and 6 black marbles from a bag that contains 70 marbles, of which 35 are white and 35 are black. If he removes an additional 15 marbles from the bag, what is the least number of these additional marbles that must be black in order for Alex to have taken more black marbles than white marbles from the bag?

(A) 11
(B) 10
(C) 9
(D) 8
(E) 7

$$(x - 6)(x - c) = x^2 - 4cx + d$$

18. In the equation above c and d are constants. If the equation is true for all values of x, what is the value of d?

(A) 42
(B) 30
(C) 24
(D) 18
(E) 12

Answers and Explanations

5. If 5 more than twice a number is equal to 20, what is 6 times the number?

    (A) 25

    (B) 30

    (C) 30 ½   Explanation:

    (D) 45 $5 + 2n = 20$

    (E) 75 $5 + 2n - 5 = 20 - 5$

$$2n = 15$$

$$\frac{2n}{2} = \frac{15}{2}$$

$$n = \frac{15}{2}$$

$$\frac{15}{2} \times 6 = \frac{90}{2} = 45$$

---

9. If $(x - y)^2 = 25$ and $(x + y)^2 = 81$, what is the value of xy?

    (A) 9

    (B) 14

    (C) 24

    (D) 27   Explanation:

    (E) 40   Step 1:

First equation: $(x - y)^2 = 25$

$$(x - y)(x - y) = 25$$

$$x^2 - 2xy + y^2 = 25$$

Step 2:

Second equation: $(x + y)^2 = 81$

$$(x + y)(x + y) = 81$$

$$x^2 + 2xy + y^2 = 81$$

Step 3:

Notice the simultaneous equations. Two or more equations with two or more variables. We want to solve for xy, so let's find a way to cancel out the x^2 and y^2.

Multiply the first equation by -1:
$$-1(x^2 - 2xy + y^2) = -1(25)$$
$$-x^2 + 2xy - y^2 = -25$$
Add the two equations together:

$$
\begin{array}{rcl}
x^2 + 2xy + y^2 &=& 81 \\
+ \quad -x^2 + 2xy - y^2 &=& -25 \\
\hline
0 + 4xy + 0 &=& 56 \\
\dfrac{4xy}{4} &=& \dfrac{56}{4} \\
xy &=& 14
\end{array}
$$

Notice what happens if we don't multiply by -1 first. We cancel the variables we want to keep, xy:

$$
\begin{array}{rcl}
x^2 - 2xy + y^2 &=& 25 \\
+ \quad x^2 + 2xy + y^2 &=& 81 \\
\hline
2x^2 \quad\quad + 2y^2 &=& 106
\end{array}
$$

---

10. If $2^{4a} \cdot 2^{2b} = 64$, and a and b are positive integers, what is the value of $2a + b$?

(A) $\frac{3}{2}$

(B) 2

(C) 3 ⟵ circled

(D) 4

(E) 6

Explanation:

Break the equation down by converting 64 to the common base of 2:

$64 = 2^6$

$2^{4a} \cdot 2^{2b} = 2^6$

$2^{4a + 2b} = 2^6$

Now get rid of the common base of 2 and follow your exponent rules by adding the exponents.

$4a + 2b = 6$

Now, divide by 2:

$$\frac{4a + 2b}{2} = \frac{4a}{2} + \frac{2b}{2} = \frac{6}{2}$$

$2a + b = 3$

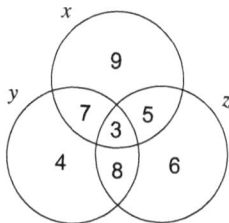

11. The Venn diagram above represents sets $x, y,$ and z where the number in each region indicates the number of elements in that region. How many elements are common to sets y and z?

(A) 33

(B) 23

(C) 12

(D) 11 ⟵ circled

(E) 8

Explanation:
8 elements are included in the overlap between y and z. 3 elements are included in the overlap between x, y and z. With your pencil, shade in the relevant overlap and then add.

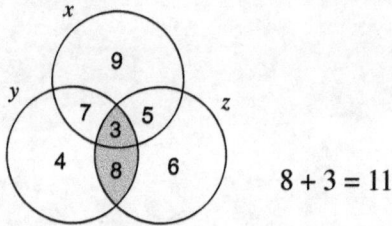

$$8 + 3 = 11$$

---

15. If $b > 1$ in the equation $b^x \cdot b^{-5} = b^{-y}$ and $b^y \div b^{-3} = b^x$, what is the value of x?

Answer: 4

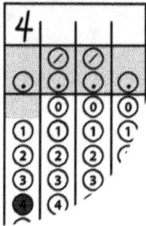

Explanation:
Notice the common bases. We can get rid of the base b and follow our exponent rules.

$x + (-5) = -y$ and $y - (-3) = x$
$x - 5 = -y$ $y + 3 = x$

Add the y and the 5   Subtract the x and 3
to both sides: from both sides:

$x + y = 5$ $-x + y = -3$

From here it's really just a simultaneous equation problem. Line up the x and y in the same position, set one equation on top of the other, and add the two equations together.

$$
\begin{array}{r}
x + y = 5 \\
+ \ -x + y = -3 \\
\hline
2y = 2 \\
\frac{2y}{2} = \frac{2}{2} \\
y = 1
\end{array}
$$

Substitute back in $y = 1$ to find x:

$$x + y = 5$$
$$y = 1$$
$$x + 1 = 5$$
$$x + 1 - 1 = 5 - 1$$
$$x = 4$$

124

16. Alex removes 7 white and 6 black marbles from a bag that contains 70 marbles, of which 35 are white and 35 are black. If he removes an additional 15 marbles from the bag, what is the least number of these additional marbles that must be black in order for Alex to have taken more black marbles than white marbles from the bag?

(A) 11
(B) 10
(C) 9
(D) 8
(E) 7

Explanation:

Do the A.C.T - Answer Choice Test

Start with the least: (E) 7 additional black marbles

7 + (additional white marbles) = 15
    additonal white marbles = 8

Original black = 6   additional black = 7
    6 + 7 = 13

Original white = 7   additonal white = 8
    8 + 7 = 15
    13 $\not>$ 15

More black marbles need to be removed. Eliminate (E).

Try (D) 8 additional black marbles

8 + (additional white) = 15
    additional white = 7

Original black = 6   additional black = 8
    6 + 8 = 14

Original white = 7   additional white = 7
    7 + 7 = 14

    14 $\not>$ 14

Eliminate (D).

Try (C) 9 additional black marbles

9 + (additional white) = 15
    additional white = 6

Original black = 6   additional black = 9
    6 + 9 = 15

Original white = 7   additional white = 6
    7 + 6 = 13
    15 > 13 ✓

(C) works!

$$(x - 6)(x - c) = x^2 - 4cx + d$$

18. In the equation above c and d are constants. If the equation is true for all values of x, what is the value of d?

(A) 42
(B) 30
(C) 24
(D) 18
(E) 12

Explanation:
The general form of a quadratic function is $y = ax^2 + bx + c$. Notice that we have an x^2 term, an x term, and a non-x term (c). When $x = 0$, $y = c$, therefore the non-x term is ALWAYS the y-intercept. When setting quadratics equal to each other, equate like terms; let's do this by expanding the left side using the FOIL method:

$$(x - 6)(x - c)$$

$$x^2 - cx - 6x + 6c = x^2 - 4cx + d$$

First, cancel the x^2 from both sides:
$$x^2 - cx - 6x + 6c - x^2 = x^2 - 4cx + d - x^2$$
$$-cx - 6x + 6c = -4cx + d$$

Now, split into two equations:
$$-cx - 6x = -4cx \quad \text{and} \quad 6c = d$$

Solve the first equation for c:
$$-cx - 6x = -4cx$$
$$-cx - 6x + cx = -4cx + cx$$
$$-6x = -3cx$$
$$\frac{-6x}{-3x} = \frac{-3cx}{-3x}$$
$$2 = c$$

Now plug $c = 2$ into the second equation:
$$6c = d$$
$$6(2) = 12$$

Now let's do a combined algebra and arithmetic drill.

Arithmetic & Algebra Drill

8. If 60 percent of Doug's DVDs cost $18 each and if 40 percent of his DVDs cost $11 each, what is the average (arithmetic mean) cost per DVD?

(A) $14.00
(B) $14.50
(C) $14.75
(D) $15.00
(E) $15.20

9. A bag of nuts contains almonds, cashews, and peanuts. The number of cashews and peanuts is twice the number of almonds. If one nut is to be chosen at random from the bag, the probability that an almond will be chosen is 5 times the probability that a cashew will be chosen. If there are 15 almonds in the bag, what is the total number of nuts in the bag?

(A) 25
(B) 30
(C) 45
(D) 51
(E) 55

NUMBER OF SEASON TICKET HOLDERS

Year	2007	2008	2009
Theatre X	120	170	250
Theatre Y	200	350	500

AVERAGE NUMBER OF PLAYS ATTENDED
BY SEASON TICKET HOLDERS AT THEATRE Y

Year	Plays
2007	4
2008	10
2009	12

9. The first table shows the number of season ticket holders at two theatres, X and Y, during the years 2007-2009. The second table shows the average (arithmetic mean) number of plays attended by each of the season ticket holders at Theatre Y during each of these years. Based on this information, which of the following is the best approximation of the total number of plays attended by season ticket holders at Theatre Y during the years 2007 – 2009?

(A) 5,180
(B) 10,300
(C) 13,000
(D) 14,040
(E) 27,300

Salaries of Certain Employees

	Month 1	Month 2
CEOs	$40,000	$43,200
Executives	$20,000	$42,260
Salespersons	$10,000	$13,790
Total	$70,000	$99,250

9. The monthly salaries for the CEOs, executives, and salespersons during a company's first two months of operation are listed in the table above. What was the average (arithmetic mean) increase in wages, in dollars, for these 3 types of employees from Month 1 to Month 2?

NUMBER OF DIAMOND NECKLACES
SOLD AT SIX COMPANIES

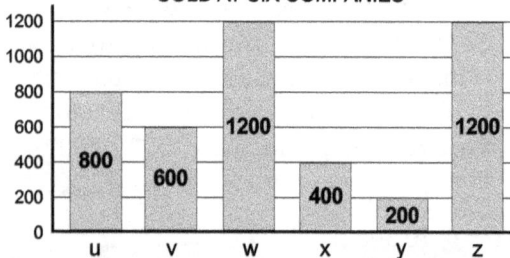

10. Based on the chart above, what was the total number of diamond necklaces sold at those jewelry stores for which the number of diamond necklaces sold was less than the mode but greater than the median number of diamond necklaces sold at the 6 companies?

10. If a positive even integer y is picked at random from the positive integers less than or equal to 12, what is the probability that $3y + 4 < 17$?

(A) $\frac{1}{4}$

(B) $\frac{1}{3}$

(C) $\frac{2}{5}$

(D) $\frac{5}{12}$

(E) $\frac{1}{2}$

Altitude (in meters)	Pressure (in hPa)
22	62.5
11	250
5.5	500
3	700
0	1000

11. The chart above shows air pressure as a function of altitude. Which of the following graphs demonstrates the relationship between air pressure and altitude?

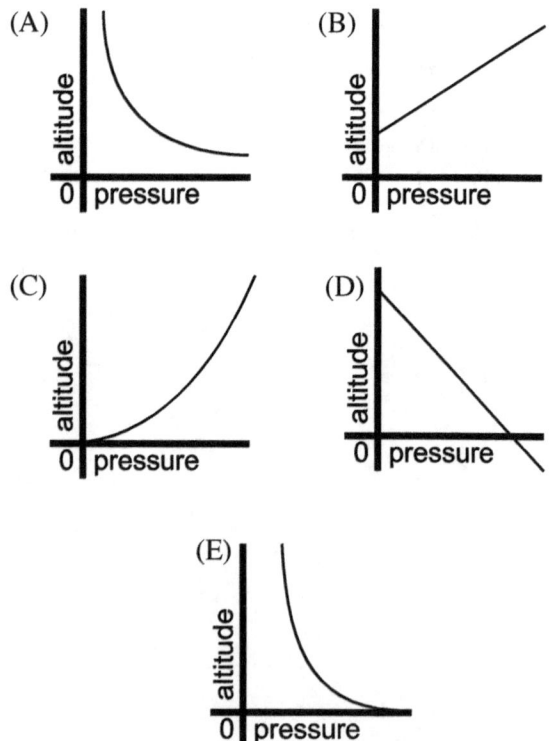

12. The number of prospective students accepted into PhD programs in the United States for the 2010 school year is approximately 2000. If 1200 students have enrolled in PhD programs so far, and 70 percent of those enrolled are male and 30 percent of those enrolled are female, how many of the remaining students to enroll must be female in order for half of the total students accepted to be female?

(A) 200
(B) 350
(C) 480
(D) 640
(E) 800

13. How many different ordered pairs (x, y) are there such that x is an odd integer, where $3 \leq x \leq 11$, and y is an integer, where $3 < y < 11$?

(A) 21
(B) 25
(C) 35
(D) 45
(E) 63

12. In the 28-day month of February, for every four days Patty was on time for school, there were three days she was tardy. The number of days Patty was on time for school is how much greater than the number of days Patty was tardy for school in the month of February?

(A) 3
(B) 4
(C) 5
(D) 6
(E) 7

$$\frac{u}{2} + \frac{t}{4} + \frac{v}{12} = 2$$

14. In the equation above, u, t, and v are distinct positive integers. What is one possible value of the product of utv?

(A) 9
(B) 15
(C) 16
(D) 18
(E) 24

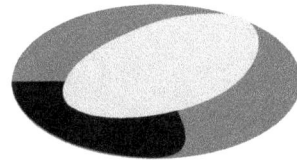

13. If $n > 0$, and 22% of n is equal to 40% of p, then 15% of n equals approximately what percent of p?

(A) 20%
(B) 27%
(C) 50%
(D) 67%
(E) It cannot be determined from the given information.

16. To gain brand name recognition, a new shoe company is designing a logo using 3 different colors according to the design above. If 6 different colors are available for the design, how many differently colored designs are possible?

(A) 10
(B) 20
(C) 25
(D) 60
(E) 120

Answers and Explanations

Answer Key:

8. (E)	**10.** 800	**12.** (D)	**13.** (C)
9. (C)	**10.** (B)	**12.** (B)	**14.** (E)
9. (B)	**11.** (E)	**13.** (B)	**16.** (E)
9. 9,750			

8. If 60 percent of Doug's DVDs cost $18 each and if 40 percent of his DVDs cost $11 each, what is the average (arithmetic mean) cost per DVD?

(A) $14.00
(B) $14.50
(C) $14.75
(D) $15.00
(E) $15.20

Explanation:
We have a Sneaky Plug In on our hands!
Suppose Doug has 10 DVDs.

60% of 10 is 6, and 6 DVDs cost $18 each.
$6 \times 18 = \$108$

40% of 10 is 4.
$4 \times 11 = \$44$

The total cost of DVDs is $108 + 44 = 152$

Our formula for average is $\dfrac{\text{total}}{\text{number of things}}$ so: $\dfrac{152}{10} = \$15.20$

9. A bag of nuts contains almonds, cashews, and peanuts. The number of cashews and peanuts is twice the number of almonds. If one nut is to be chosen at random from the bag, the probability that an almond will be chosen is 5 times the probability that a cashew will be chosen. If there are 15 almonds in the bag, what is the total number of nuts in the bag?

(A) 25
(B) 30
(C) 45
(D) 51
(E) 55

Explanation:
Let's make Almonds = A, Cashews = C and Peanuts = P

"number of cashews and peanuts is twice the number of almonds"
$$C + P = 2(A)$$

The question tells us there are 15 almonds in the bag.
$A = 15$, so
$C + P = 2(A) = 2(15) = 30$
$A + C + P = \text{Total}$
$15 + 30 = 45 \text{ Total}$

Or, we could use a ratio grid:

	Almonds	Cashews	Peanuts	Total
Ratio	5	1	9	15
Multiplier	× 3	× 3	× 3	× 3
Actual #	15	3	27	45

$$3 + P = 2(15)$$
$$3 + P = 30$$
$$P = 27$$

$$15 + 3 + 27 = 45$$

NUMBER OF SEASON TICKET HOLDERS

Year	2007	2008	2009
Theatre X	120	170	250
Theatre Y	200	350	500

AVERAGE NUMBER OF PLAYS ATTENDED
BY SEASON TICKET HOLDERS AT THEATRE Y

Year	Plays
2007	4
2008	10
2009	12

9. The first table shows the number of season ticket holders at two theatres, X and Y, during the years 2007-2009. The second table shows the average (arithmetic mean) number of plays attended by each of the season ticket holders at Theatre Y during each of these years. Based on this information, which of the following is the best approximation of the total number of plays attended by season ticket holders at Theatre Y during the years 2007 – 2009?

(A) 5,180
(B) 10,300
(C) 13,000
(D) 14,040
(E) 27,300

Explanation:
For each year, multiply the number of ticket holders by the average number of plays attended:
$$200 \times 4 = 800$$
$$350 \times 10 = 3500$$
$$500 \times 12 = 6000$$
Add your yearly approximations to get your total approximation:
$$800 + 3500 + 6,000 = 10,300$$

Salaries of Certain Employees

	Month 1	Month 2
CEOs	$40,000	$43,200
Executives	$20,000	$42,260
Salespersons	$10,000	$13,790
Total	$70,000	$99,250

9. The monthly salaries for the CEOs, executives, and salespersons of Company A for the first two months of operation are listed in the table above. What was the average (arithmetic mean) increase in wages, in dollars, for these 3 types of employees from Month 1 to Month 2?

Explanation:
In order to find the increase, find the difference between the 2 monthly totals.

$$\begin{array}{r} 99,250 \\ -\ 70,000 \\ \hline 29,250 \end{array}$$

To find the average increase, put the total increase (29,250) over the number of types of employees (3).

$$\frac{29,250}{3} = 9,750$$

NUMBER OF DIAMOND NECKLACES
SOLD AT SIX COMPANIES

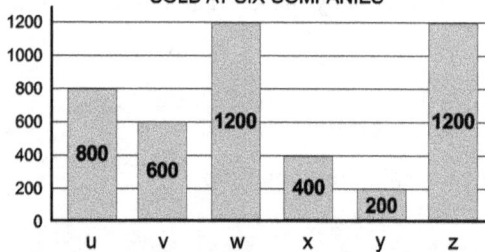

10. Based on the chart above, what was the total number of diamond necklaces sold at those jewelry stores for which the number of diamond necklaces sold was less than the mode but greater than the median number of diamond necklaces sold at the 6 companies?

Explanation:

Step 1:

First, find the median. Remember to arrange the numbers in order from least to greatest.

~~200~~, ~~400~~ (600, 800) ~~1200~~, ~~1200~~

$$\frac{800 + 600}{2} = \frac{1400}{2} = 700$$

Median = 700

Step 2:

Now determine the mode. Mode is the number that occurs most frequently. The only number that occurs more than any other is 1200 so, the mode = 1200

$$700 < x < 1200$$

Company u's total is the only one that falls within this range.

so $x = 800$

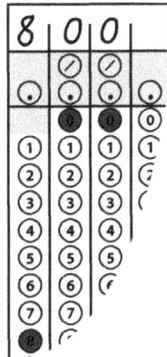

---

10. If a positive even integer y is picked at random from the positive integers less than or equal to 12, what is the probability that $3y + 4 < 17$?

(A) $\frac{1}{4}$

(B) $\frac{1}{3}$

(C) $\frac{2}{5}$

(D) $\frac{5}{12}$

(E) $\frac{1}{2}$

Explanation:
Write out all your possibilities:
2, 4, 6, 8, 10, 12

Now, Plug in to the equation: $3y + 4 \leq 17$

$3(2) + 4 \leq 17$ $3(4) + 4 \leq 17$ $3(6) + 4 \leq 17$ $3(8) + 4 \leq 17$
$\quad 10 \leq 17$ ✓ $\quad 16 \leq 17$ ✓ $\quad 22 \leq 17$ ✗ $\quad 28 \leq 17$ ✗

$3(10) + 4 \leq 17$ $3(12) + 4 \leq 17$
$\quad 34 \leq 17$ ✗ $\quad 40 \leq 17$ ✗

2 and 4 meet the criterion, so two out of six possibilities:

$$\frac{2}{6} = \frac{1}{3}$$

It can also be solved algebraically. Solve for y:

$$3y + 4 \le 17$$
$$3y + 4 - 4 \le 17 - 4$$
$$3y \le 13$$
$$\frac{3y}{3} \le \frac{13}{3}$$
$$y \le \frac{13}{3}$$
$$y \le 4\frac{1}{3}$$

Of the 6 possibilities, only 2 and 4 are less than or equal to $4\frac{1}{3}$. So, $\frac{2}{6}$ or $\frac{1}{3}$ fit the criterion.

Altitude (in meters)	Pressure (in hPa)
22	62.5
11	250
5.5	500
3	700
0	1000

11. The chart above shows air pressure as a function of altitude. Which of the following graphs demonstrates the relationship between air pressure and altitude?

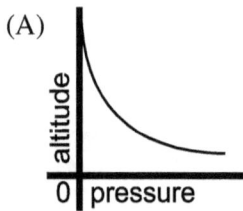

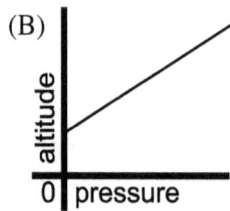

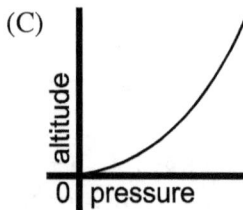

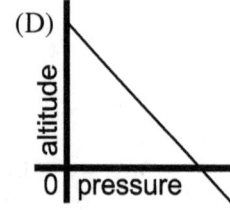

(A)

(B)

(C)

(D)

(E)

Explanation:
Make a graph using the given points:

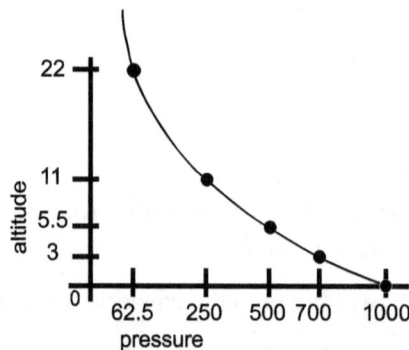

The only answer choices that have similar shapes to the graph we made are (A) and (E). But (A) doesn't contain the point $(1000, 0)$, so the answer is (E). You could also use the answer choices to help you ballpark the right answer. Altitude and pressure have an inverse relationship (as altitude increases, pressure decreases), so cancel (B) and (C). There is no constant slope and the function is not linear, so cancel (D).

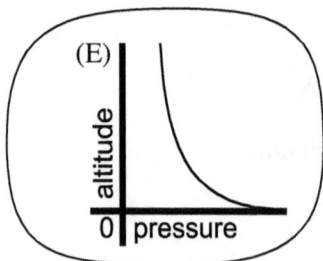

12. The number of prospective students accepted
into PhD programs in the United States for
the 2010 school year is approximately 2000. If
1200 students have enrolled in PhD programs
so far, and 70 percent of those enrolled are
male and 30 percent of those enrolled are
female, how many of the remaining students to
enroll must be female in order for half of the
total students accepted to be female?

(A) 200
(B) 350
(C) 480
(D) 640
(E) 800

Explanation:
Don't get lost in the words. Organize your information.
2000 accepted, 1,200 total currently enrolled
"30 percent of those enrolled are female"

Translate: $\frac{30}{100} \cdot 1200 = 360$ females currently enrolled

Half of the total are female:

$2000 \cdot \frac{1}{2} = 1000$

$1000 - 360 = 640$ females left to enroll

---

12. In the 28-day month of February, for every
four days Patty was on time for school, there
were three days she was tardy. The number of
days Patty was on time for school is how much
greater than the number of days Patty was
tardy for school in the month of February?

(A) 3
(B) 4
(C) 5
(D) 6
(E) 7

Explanation:
Set up a ratio grid with the given information:

	Days on time	Days tardy	Total
Ratio	4	3	7
Multiplier			
# of days			28

Ask: What do I multiply 7 by to get 28?

Now fill in the missing pieces:

	Days on time	Days tardy	Total
Ratio	4	3	7
Multiplier	× 4	× 4	× 4
# of days	16	12	28

So Patty was on time 16 days out of the month and tardy 12 days and $16 - 12 = 4$.

13. If $n > 0$, and 22% of n is equal to 40% of p, then 15% of n equals approximately what percent of p?

(A) 20%
(B) 27%
(C) 50%
(D) 67%
(E) It cannot be determined
    from the given information.

Explanation:

Answer choice (E) is rarely the answer and is only correct if you can prove that more than one or none of the other choices work. Otherwise, it's there to distract you. Cancel it! Notice percents in the answer choices which is a tip off that it's a Sneaky Plug In.

22% of n is equal to 40% of p

$$\frac{22}{100}n = \frac{40}{100}p$$

Remember – it's always easiest to plug in 100 so we don't have to worry about converting to percents at the end.

$$n = 100$$

$$\frac{22}{100} \times 100 = \frac{40}{100}p$$

$$\frac{22}{100} \times \frac{100}{1} = \frac{40}{100}p$$

$$\frac{22}{1} \times \frac{40p}{100}$$

$$40p = 2{,}200$$

$$\frac{40p}{40} = \frac{2{,}200}{40}$$

$$p = 55$$

"15% of n equals approximately what percent of p." Remember the words "What percent" always translate to $\frac{x}{100}$.

$$\frac{15}{100} \times 100 = \frac{x}{100} \times 55$$

$$\frac{15}{100} \times \frac{100}{1} = \frac{55x}{100}$$

$$\frac{15}{1} = \frac{55x}{100}$$

$$55x = 1{,}500$$

$$\frac{55x}{55} = \frac{1{,}500}{55}$$

$$x = 27.272727\ldots$$

13. How many different ordered pairs (x, y) are there such that x is an odd integer, where $3 \le x \le 11$, and y is an integer, where $3 < y < 11$?

(A) 21
(B) 25
(C) 35
(D) 45
(E) 63

Explanation:
This is a multiple source problem. So list your possible x, y values and then set up your categories:

x	y
3	4
5	5
7	6
9	7
11	8
	9
	10

$5 \ \times \ 7 = 35$

$$\frac{u}{2} + \frac{t}{4} + \frac{v}{12} = 2$$

14. In the equation above, u, t, and v are distinct positive integers. What is one possible value of the product of utv?

(A) 9
(B) 15
(C) 16
(D) 18
(E) 24

Explanation:
The easiest thing to do first is to add those fractions together and simplify. We need to find a common denominator:

$$\frac{u}{2} \cdot \frac{6}{6} = \frac{6u}{12}$$

$$\frac{t}{4} \cdot \frac{3}{3} = \frac{3t}{12}$$

$$\frac{v}{12} \cdot \frac{1}{1} = \frac{v}{12}$$

Now add the fractions:

$$\frac{6u}{12} + \frac{3t}{12} + \frac{v}{12} = 2$$

$$\frac{6u + 3t + v}{12} = \frac{2}{1}$$

Cross multiply:

$$\frac{6u + 3t + v}{12} \diagdown \frac{2}{1}$$

$6u + 3t + v = 24$

Now we can plug in values for u, t, and v that satisfy $6u + 3t + v = 24$

Let $u = 1, t = 2, v = 12$

$6(1) + 3(2) + 12 = 24$

$6 + 6 + 12 = 24$

Now, multiply utv:

$1 \times 2 \times 12 = 24$ ✓

16. To gain brand name recognition, a new shoe company is designing a logo using 3 different colors according to the design above. If 6 different colors are available for the design, how many differently colored designs are possible?

(A) 10
(B) 20
(C) 25
(D) 60
(E) 120

Explanation:

Single source problem - order matters, because if we switch the position of the color, we would get an entirely new design.

Color 1		Color 2		Color 3		
6	×	5	×	4	=	120

Answer: (E) 120

Chapter 9
Geometry Lesson

The Geometry on the SAT is pretty basic – it's just a matter of knowing a few formulas, rules, and degree measures and then it comes down to putting the pieces of the puzzle together correctly.

Reference Information

$A = \pi r^2$
$C = 2\pi r$
$A = lw$
$A = \frac{1}{2}bh$
$V = lwh$
$V = \pi r^2 h$
$c^2 = a^2 + b^2$
Special Right Triangles

The number of degrees of arc in a circle is 360.
The sum of the measures in degrees of the angles of a triangle is 180.

The reference information box above appears at the beginning of every math section, so don't panic if you forget a formula. That being said, 25 minutes per section is not very much time, so it is important to know your geometry rules and know them well so you don't have to take the time to go searching the reference box on every geometry question you hit.

Let's start with some basic tips to tackling the geometry on the SAT.

Tip #1: A bunch of words and no figure to help you out? *Draw the figure on your own*!

It's easy to approximate if you draw a precise and accurate figure. Let's see how drawing our own figure can help us out.

9. In the coordinate plane, the points R $(-2, 2)$, S $(2, 6)$, and T $(6, 2)$ lie on a circle with center V. What are the coordinates of point V?

(A) $(0, 0)$
(B) $(2, -3)$
(C) $(3, 2)$
(D) $(2, 2)$
(E) $(1.5, 2)$

Explanation:
Draw the figure on a coordinate graph:

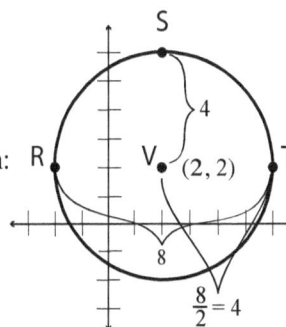

The points R and T are directly across from each other and a distance of 8 spaces apart. Half of 8 is 4 which puts point V at the 2 mark for x and the 2 mark for y.

Answer: (D) $(2, 2)$

Tip #2: *Never assume anything when a figure is "not drawn to scale"!*

How do you know if a figure is drawn to scale? It won't say "Note: figure not drawn to scale." Let's take a look at how ETS loves to play on our assumptions.

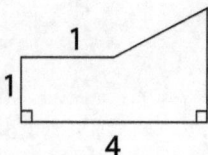

Note: Figure not drawn to scale

14. If the area of the figure above is 10, what is the perimeter of the figure?

(A) 16
(B) 17
(C) 18
(D) 19
(E) 20

Explanation:

The figure isn't drawn to scale, so we must trust what we read, more than what we see. Divide the figure into a triangle and rectangle: Solve for the area of the rectangle:

$lw = $ Area

$4 \times 1 = 4$

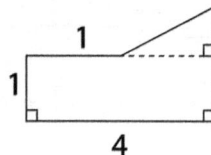

We can subtract the length of the rectangle from the segment on the top of the rectangle to find the base of the triangle: $4 - 1 = 3$

We also know that the area of the triangle is 6, since the area of the rectangle is 4 and the total area of the figure is 10. Given a base of 3 and an area of 6, we can solve for the height of the triangle:

$$\text{Area} = \frac{1}{2} bh$$

$$6 = \frac{1}{2} 3h$$

$$\frac{6}{1} \times \frac{3h}{2}$$

$$3h = 12$$

$$\frac{3h}{3} = \frac{12}{3}$$

$$h = 4$$

So, the triangle is a 3-4-5 triangle and the hypotenuse is 5.

Add the sides: $1 + 1 + 5 + 4 + 1 + 4 = 16$

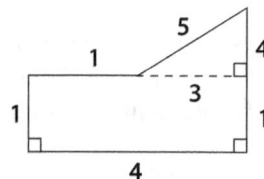

Answer: (A) 16

Tip #3: *Mark up your figure! Add all information from the given question to the given figure.*

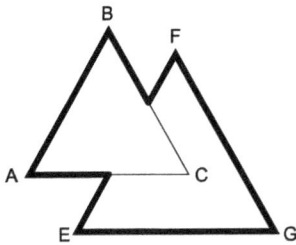

12. The figure above is made up of two equilateral overlapping triangles. If $\overline{AB} \parallel \overline{EF}$, $\overline{EF}$ bisects $\overline{BC}$, and $\overline{AB} = 16$ inches and $\overline{FG} = 24$ inches, what is the length, in inches, of the perimeter shown in bold.

(A) 72
(B) 92
(C) 96
(D) 106
(E) 120

Explanation:

If $\overline{AB}$ is 16, you know every side of $\triangle ABC$ equals 16, because the triangles are equilateral. Likewise, every side of $\triangle EFG$ is 24.

Let's label the dotted line (the bisected section of $\overline{EF}$) XY:

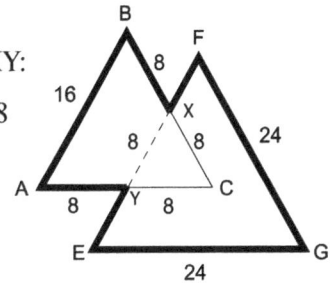

Since $\overline{EF}$ bisects $\overline{BC}$ and $\overline{AB} \parallel \overline{EF}$, we know that $\overline{XY} = 8$ and, therefore, $\overline{EF} - \overline{XY} = 24 - 8 = 16$

$24 + 24 + 16 + 16 + 8 + 8 = 96$

Answer: (C) 96

Tip #4: *When given the value of a known formula, set the value equal to the formula and solve for the unknown.*

For instance, if a problem says "the area of Circle P is 16π," immediately set $\pi r^2 = 16\pi$ and solve for the radius:

$$\pi r^2 = 16\pi$$
$$\frac{\pi r^2}{\pi} = \frac{16\pi}{\pi}$$
$$\frac{\cancel{\pi} r^2}{\cancel{\pi}} = \frac{16\cancel{\pi}}{\cancel{\pi}}$$
$$r^2 = 16$$
$$\sqrt{r^2} = \sqrt{16}$$
$$r = 4$$

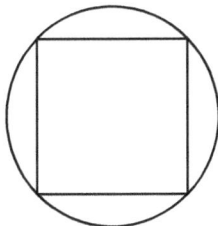

9. In the figure above, a square is inscribed in a circle whose area is 36π. What is the area of the square?

(A) 46π
(B) 46
(C) 72
(D) 27π
(E) 18

141

Explanation:

Area of circle = πr^2

$\pi r^2 = 36\pi$

$\dfrac{\pi r^2}{\pi} = \dfrac{36\pi}{\pi}$

$\dfrac{\cancel{\pi} r^2}{\cancel{\pi}} = \dfrac{36\cancel{\pi}}{\cancel{\pi}}$

$r^2 = 36$

$\sqrt{r^2} = \sqrt{36}$

$r = 6$

So, diameter = 12

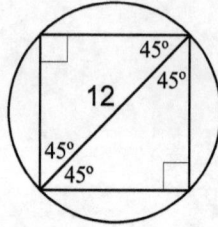

The diameter of the circle is equal to the diagonal of the square.

The diagonal divides the square into two isosceles right triangles. Each leg of an isosceles right triangle is equal to the hypotenuse divided by $\sqrt{2}$.

So, one side of the the square = $\dfrac{12}{\sqrt{2}}$

Area of square = (side)²

$\left(\dfrac{12}{\sqrt{2}}\right)^2 = \dfrac{12}{\sqrt{2}} \times \dfrac{12}{\sqrt{2}} = \dfrac{144}{2} = 72$

Answer: (C) 72

LINES AND ANGLES

Lines

1. How many degrees are in a *straight angle*?   <u>180</u>

Keep in mind that ETS might use the terminology "straight angle." Straight angle is just a fancy way to refer to a line.

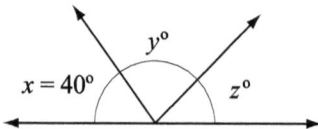

2. What is the value of $y + z$?

$$x + y + z = 180$$
$$x = 40$$
$$40 + y + z = 180$$
$$40 + y + z - 40 = 180 - 40$$
$$y + z = 140$$

Line: $\overleftrightarrow{AB}$

ETS will notate a line with a double arrow over it. Note that a line continues forever in both directions.

Segment: $\overline{AB}$

If you see AB with a line and no arrows over it, ETS is denoting a line segment with two endpoints.

Ray: $\overrightarrow{AB}$

If you see AB with a line with only one arrow on top, ETS is denoting a ray, which has one endpoint and continues forever in the direction of the arrow.

142

Angles

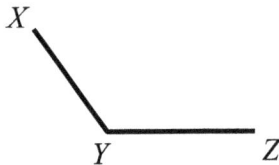

3. What is the *vertex* of angle XYZ?   <u>Y</u>

ETS will denote an angle with the angle symbol (∠). The angle above can be referred to as ∠XYZ, ∠ZYX, or simply ∠Y.

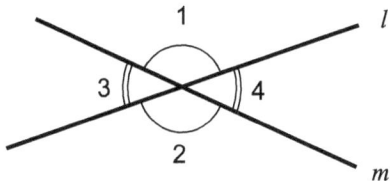

4. 1 and 2 are equal because they are what type of angles?   *<u>Vertical</u>* (They are across the vertex from each other.)

5. What is an angle less than 90° called?   *<u>Acute</u>*

6. What is an angle more than 90° called?   *<u>Obtuse</u>*

7. What is an angle equal to 90° called?   *<u>A right angle</u>*

Right angles are formed when two lines are *perpendicular* to each other.

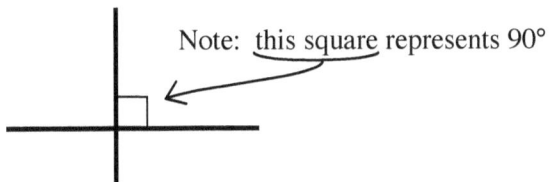

Note: this square represents 90°

The symbol for perpendicular is ⊥

8. What is the symbol for two lines that are parallel to each other? ||

Remember: parallel lines never intersect.

ETS loves to test the concept of *parallel lines cut by a transversal*. In school they made you memorize the terminology corresponding angles, alternate interior angles, exterior angles, blah, blah, blah. On the SAT, we only need to use two names for our angles: Big and Small.

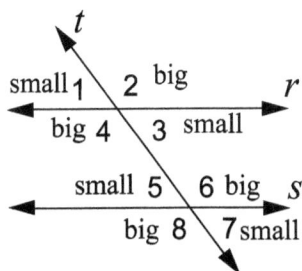

Parallel lines *r* and *s* are cut by the transversal *t*.

Now all we have to know is that the Bigs are all equal and the Smalls are all equal, and a Big plus a Small is equal to 180°.

Let's say ∠2 is 120°. Because ∠1 and ∠2 sit on a straight line, ∠1 is 60° – from there we can fill in the rest of the degree measures – the big angles are each 120° and the small angles are each 60°.

Let's try an SAT problem that tests this concept:

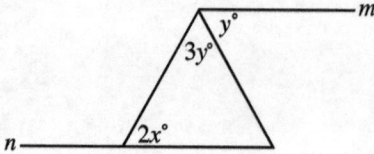

7. In the figure shown above, lines m and n are parallel and $x = 16$. What it the value of y?

(A) 33
(B) 34
(C) 35
(D) 36
(E) 37

Explanation:
Extend m and n to see that the third angle of the triangle equals y, since they are both small angles.

Since the sum of a triangle's angles is 180,
$2x + 3y + y = 180$
$2x + 4y = 180$
$2(16) + 4y = 180$
$32 + 4y = 180$
$32 + 4y - 32 = 180 - 32$
$4y = 148$
$\dfrac{4y}{4} = \dfrac{148}{4}$
$y = 37$

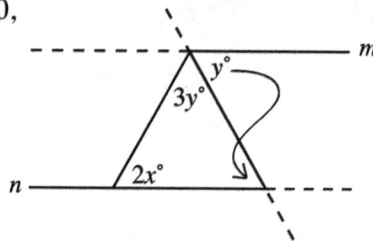

Answer: (E) 37

We could also use A.C.T and plug in our answer choices for y, starting with (C) 35:
(C) 35
$2x + 4y = 180$
$2(16) + 4(35) =$
$2 + 140 = 172$

We need a bigger number, so jump to (E):

(E) 37
$2x + 4y = 180$
$2(16) + 4(37) =$
$32 + 148 = 180$ ✓

TRIANGLES

1. How many *degrees* are in a triangle?   180

2. What is the formula for *area* of a triangle? $\frac{1}{2}bh$

3. What is a triangle with all equal sides and all equal angles called?   *Equilateral*

4. What is the degree measure of each of the angles in an equilateral triangle?   60

5. What is a triangle with two equal sides and two equal angles called? *__Isosceles__*

6. What is this figure called? *__A right triangle__*

a c b

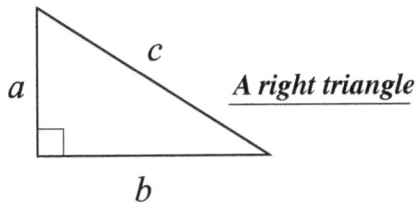

7. What is c called? *__The hypotenuse__*

8. What are a and b called? *__The legs__*

Remember: the legs are always attached to the right angle of a right triangle, and the longest side of any triangle is always opposite the largest angle.

9. When given the measurements of two sides of a right triangle, how do you find the measurement of the third side? *__The Pythagorean Theorem__*: $a^2 + b^2 = c^2$

Let's practice a bit with the Pythagorean Theorem:

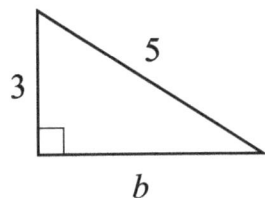

3 5 b

Find b.

Explanation:
$$a^2 + b^2 = c^2$$
$$3^2 + b^2 = 5^2$$
$$9 + b^2 = 25$$
$$9 + b^2 - 9 = 25 - 9$$
$$b^2 = 16$$
$$\sqrt{b^2} = \sqrt{16}$$
$$b = 4$$

Be careful! c is the hypotenuse, and the hypotenuse is always the longest side of a right triangle.

Special Triangles

A **3 – 4 – 5** right triangle is a special type of triangle called a *Pythagorean Triple*. Other Triples are **6 – 8 – 10** and **5 – 12 – 13**. Many of the triangles on the SAT are special triangles. Memorize them and learn to spot them to save time!

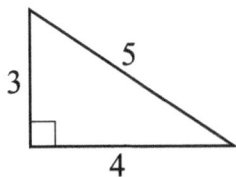

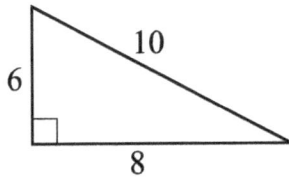

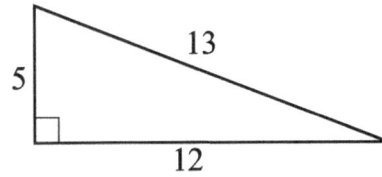

3 5 4   6 10 8   5 13 12

But what happens when they only give you one side of a right triangle? Well, if it's a **45 – 45 – 90** triangle or a **30 – 60 – 90** triangle you're in luck!

You can always solve for the two unknown sides of a **45 – 45 – 90** or **30 – 60 – 90** triangle as long as you know your rules. If you forget, just check the reference box at the beginning of the section.

Let's start with a **45 − 45 − 90**:

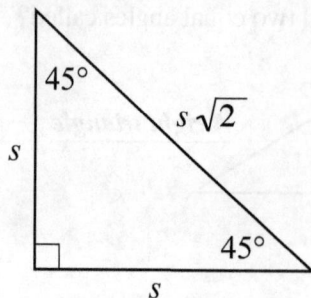

s represents the legs of the triangle and $s\sqrt{2}$ represents the hypotenuse. Whenever we need to go from a leg to the hypotenuse simply multiply by $\sqrt{2}$ and when going from hypotenuse to leg, divide by $\sqrt{2}$.

Here is a chart to help you memorize the rules:

45 − 45 − 90	
WHEN GOING FROM	ALWAYS
Leg to Hypotenuse	Multiply by $\sqrt{2}$
Hypotenuse to Leg	Divide by $\sqrt{2}$

Let's try a few:

Leg to Hypotenuse

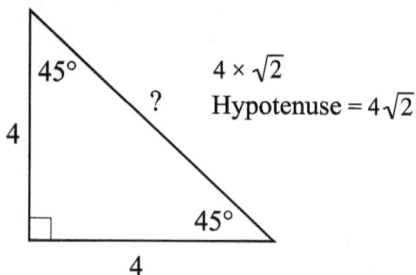

$4 \times \sqrt{2}$
Hypotenuse $= 4\sqrt{2}$

Hypotenuse to Leg

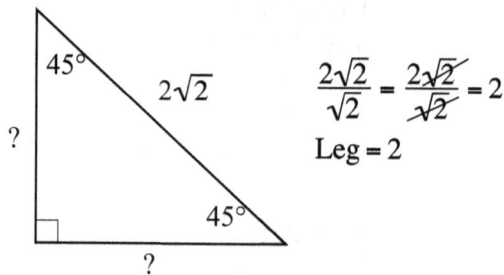

$\dfrac{2\sqrt{2}}{\sqrt{2}} = \dfrac{2\cancel{\sqrt{2}}}{\cancel{\sqrt{2}}} = 2$

Leg $= 2$

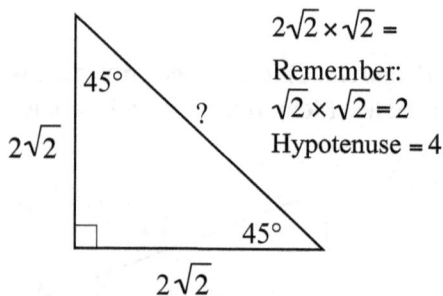

$2\sqrt{2} \times \sqrt{2} =$
Remember:
$\sqrt{2} \times \sqrt{2} = 2$
Hypotenuse $= 4$

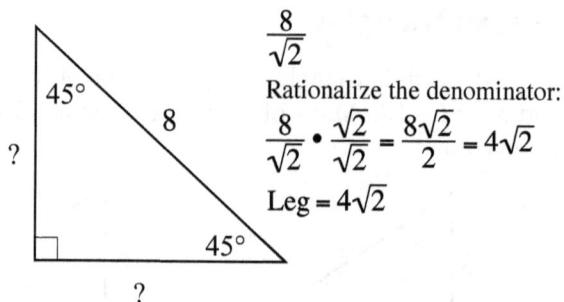

$\dfrac{8}{\sqrt{2}}$

Rationalize the denominator:

$\dfrac{8}{\sqrt{2}} \cdot \dfrac{\sqrt{2}}{\sqrt{2}} = \dfrac{8\sqrt{2}}{2} = 4\sqrt{2}$

Leg $= 4\sqrt{2}$

The only other thing to know about 45 − 45 − 90 triangles is that when you cut a square along its diagonal you form two 45 − 45 − 90 triangles.

146

For example:

The diagonal of square ABCD is 6. What is the area of the square?

Explanation:
Divide the square into two $45-45-90$ triangles:

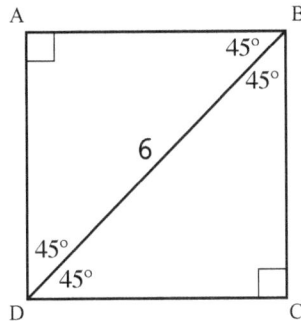

The hypotenuse is 6, so we divide by $\sqrt{2}$ to find the sides:

$$s = \frac{6}{\sqrt{2}}$$

The area of a square is s^2:

$$Area = s^2 = \left(\frac{6}{\sqrt{2}}\right)^2$$

$$= \left(\frac{6}{\sqrt{2}}\right)\left(\frac{6}{\sqrt{2}}\right) = \frac{36}{2} = 18$$

Our next special triangle is the **$30-60-90$**.

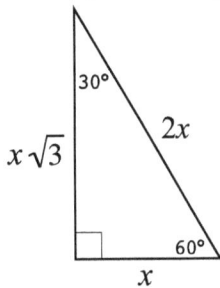

The first difference to note is the $\sqrt{3}$. A good way to remember the difference between these two types of triangles is to think: a 30–60–90 has 3 different angles $\longrightarrow \sqrt{3}$ and a 45–45–90 has two different angles $\longrightarrow \sqrt{2}$! Be careful, ETS loves to complicate things further by twisting and turning its triangles. The best way to get around this is to start with the smallest angle.

Spot your 30° angle first. The side across from the 30° is the **short side**, because the shortest side of a triangle is always across from the smallest angle. The short side of a 30–60–90 is x. The side across from the 60° angle is the **medium side**, because the medium side of a triangle is always across from the middle angle measure. The medium side is $x\sqrt{3}$. And the **long side** (or hypotenuse) is across from the 90° angle, because the biggest side of a triangle is always across from the biggest angle. The biggest side is $2x$. No matter what side ETS gives you to start, you should always find your short side. If given the short side, multiple by 2 to find the long side, and multiply by $\sqrt{3}$ to find the medium side. If given the long side, divide by 2 to find the short side and then multiply by $\sqrt{3}$ to find the medium side. If given the medium side, divide by $\sqrt{3}$ to find the short side and then multiply by 2 to find the long side.

Here is a chart to make the rules simpler:

30 – 60 – 90	
WHEN GOING FROM	ALWAYS
Short to Long	Multiply by 2
Short to Medium	Multiply by $\sqrt{3}$
Long to Short	Divide by 2
Medium to Short	Divide by $\sqrt{3}$

Let's try some. Remember to rationalize and simplify those square roots!

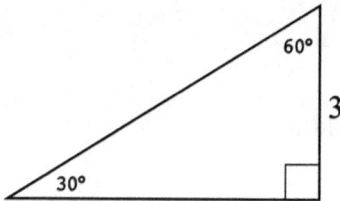

Short = 3
$3 \times 2 = 6$
Long = 6
$3 \times \sqrt{3}$
Medium = $3\sqrt{3}$

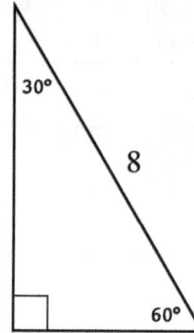

Long = 8
$\frac{8}{2} = 4$
Short = 4
$4 \times \sqrt{3} = 4\sqrt{3}$
Medium = $4\sqrt{3}$

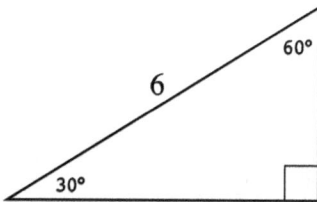

Long = 6
$\frac{6}{2} = 3$
Short = 3
$3 \times \sqrt{3} = 3\sqrt{3}$
Medium = $3\sqrt{3}$

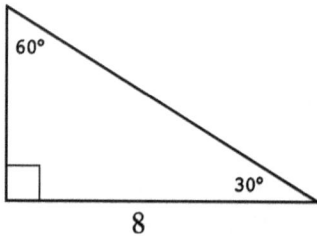

Medium = 8
$\frac{8}{\sqrt{3}} \times \frac{\sqrt{3}}{\sqrt{3}} = \frac{8\sqrt{3}}{3}$
Short = $\frac{8\sqrt{3}}{3}$
$\frac{8\sqrt{3}}{3} \times 2 = \frac{16\sqrt{3}}{3}$
Long = $\frac{16\sqrt{3}}{3}$

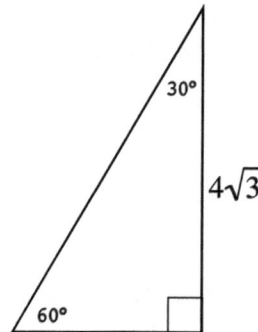

Medium = $4\sqrt{3}$
$\frac{4\sqrt{3}}{\sqrt{3}} = \frac{4\sqrt{3}}{\sqrt{3}} = 4$
Short = 4
$4 \times 2 = 8$
Long = 8

You should know that when you cut a rectangle along its diagonal you do not necessarily form two 30 – 60 – 90 triangles, so don't make any assumptions. What you can depend on, however, is that **when you bisect an equilateral triangle you form two 30 – 60 – 90 triangles.**

Let's say ETS asks, "Triangle ABC is an equilateral triangle with side of 6 in length. What is the area of the triangle?"

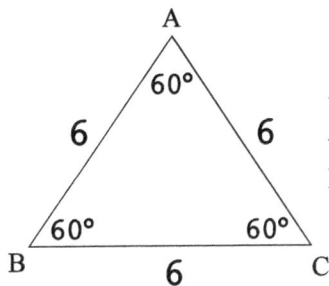

We can divide this triangle so we have two $30-60-90$ triangles with a short side of 3 so we can solve for the height (the medium side): Short = 3

Medium = $3\sqrt{3}$

$h = 3\sqrt{3}$ $b = 6$

$Area = \frac{1}{2}bh$

$= \frac{1}{2} \times 6 \times 3\sqrt{3}$

$= 9\sqrt{3}$

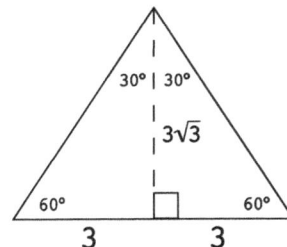

Remember to use the WHOLE base!

Now, let's look at a harder application of these concepts. Give it a shot before reading the explanation!

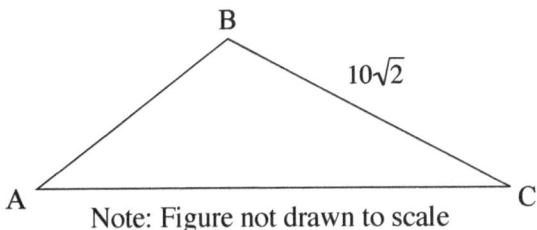

Note: Figure not drawn to scale

17. In $\triangle ABC$, the measure of $\angle C$ is $45°$ and the measure of $\angle A$ is $60°$. What is the length of segment $\overline{AC}$?

(A) 20

(B) $10\sqrt{2} + 10$

(C) $10\sqrt{3} + 10$

(D) $\frac{10\sqrt{3}}{3} + 10$

(E) $10\sqrt{3} + 20$

Explanation:
Note: this is a difficult problem (#17!) and is made harder because we cannot fully trust the drawing, which is not drawn to scale. Remember, a harder problem on the SAT usually just means more work and often with geometry, they are looking for shapes within the given shapes. So, given 45° and 60°, we are looking for multiple triangles.

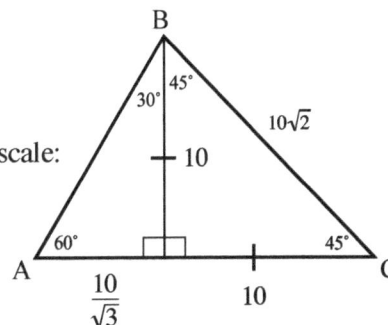

Step 1:
Divide the triangle into two right triangles.
One a 30-60-90 and the other a 45-45-90.
Sometimes it helps to redraw the figure more to scale:

Step 2:

BC is the hypotenuse of our $45-45-90$ triangle. To get from the hypotenuse to the legs, divide $10\sqrt{2}$ by $\sqrt{2}$:

$\frac{10\sqrt{2}}{\sqrt{2}} = 10$

Step 3:
Now we have the medium side of our 3-60-90 triangle. We need our short side, so divide by $\sqrt{3}$:

$\frac{10}{\sqrt{3}}$

Step 4:

Before we add, we need to multiply the square root out of the denominator:

$$\frac{10}{\sqrt{3}} \cdot \frac{\sqrt{3}}{\sqrt{3}} = \frac{10\sqrt{3}}{3}$$

Add our two values to get the length of $\overline{AC}$.

$$\overline{AC} = \frac{10\sqrt{3}}{3} + 10$$

Answer: (D) $\frac{10\sqrt{3}}{3} + 10$

What about finding the lengths of the sides of a triangle that isn't a right triangle? If ETS gives you two sides of a triangle that is not a right triangle they will not ask you to find the value of the third side, they will only ask you to find a possible value of the third side, because that's all you can do. In cases such as these, ETS is testing the **THIRD SIDE TRIANGLE RULE.**

Third Side Triangle Rule: The sum of the lengths of any two sides of a triangle must be greater than the length of the third side.

Let's say ETS says, "Triangle ABC has sides of lengths 6 and 9. What is a possible length of the third side?" Find the sum of the two given sides and the difference of the two given sides and the third side has to fall somewhere in between.

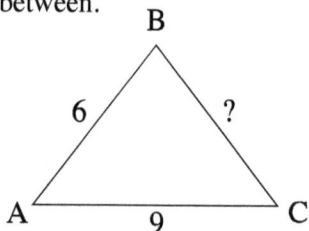

Note: Figure not drawn to scale

$$9 + 6 = 15$$
$$9 - 6 = 3$$
$$3 < x < 15$$

Let's see how this works on an SAT problem:

8. Two sides of a triangle each have length 7.
All of the following could be the length of the third side EXCEPT

(A) 14
(B) 12
(C) $\sqrt{60}$
(D) 5
(E) 1

Explanation:
When looking for the possibilities for the third side of a triangle find the sum and the difference and the third side has to be greater than the difference and less than the sum:

$$7 + 7 = 14$$
$$7 - 7 = 0$$
$$0 < x < 14$$

The third side must be less than 14, not less than or equal to, so (A) is our EXCEPT.

Answer: (A) 14

150

Similar Triangles: Triangles that have equal angles have sides that are proportional in length.

Let's try a couple of examples:

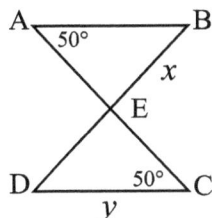

Note: Figure not drawn to scale

10. In the figure, $\overline{AC}$ and $\overline{BD}$ intersect at point E. If $\overline{AB} = 4$, $\overline{BE}$ is equal to x, and the value of $\overline{DC}$ is equal to y, what is the value of $\overline{DE}$?

(A) x

(B) xy

(C) $\dfrac{x^2y}{4}$

(D) x^2

(E) $\dfrac{xy}{4}$

Explanation:

See those variables in the answer choices - this is a Basic Plug In! The two triangles must be similar because all of the angles are equal.

Plug In $y = 8$ and $x = 6$:

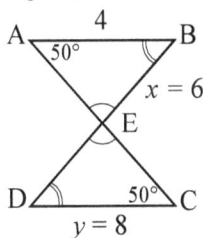

Set up the proportion: $\dfrac{AB}{BE} = \dfrac{y}{DE}$

Plug in values: $\dfrac{4}{6} = \dfrac{8}{DE}$

Cross multiply: $4DE = 48$

$\dfrac{4DE}{4} = \dfrac{48}{4}$

$DE = 12$ $\boxed{12}$

Now plug in our supposed x and y to the answer choices to find a match for our boxed value:

(A) x
  6

(B) xy
  $6 \times 8 = 48$

(C) $\dfrac{x^2y}{4}$
  $\dfrac{6^2 \times 8}{4}$
  $\dfrac{36 \times 8}{4}$
  $\dfrac{288}{4} = 72$

(D) x^2
  $6^2 = 36$

(E) $\dfrac{xy}{4}$
  $\dfrac{6 \times 8}{4}$
  $\dfrac{48}{4} = 12$

Answer: (E) $\dfrac{xy}{4}$

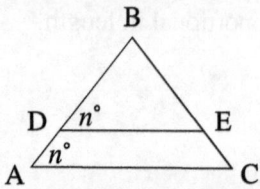

11. $\overline{BE}$ is equal to two thirds of $\overline{BC}$ in the figure above. If $\overline{BC}$ is equal to 12, what is the value of $\frac{DE}{AC}$?

(A) $\frac{1}{4}$

(B) $\frac{1}{3}$

(C) $\frac{1}{2}$

(D) $\frac{2}{3}$

(E) $\frac{3}{2}$

Explanation:

The shape of the figure is a big hint. The two triangles have $\angle B$ in common as well as an angle of $n°$ in common, so they must be similar triangles. All corresponding sides are proportional in length. Since $\overline{BE}$ is two thirds of $\overline{BC}$, $\overline{DE}$ is also two thirds of $\overline{AC}$. Note: parallel lines cutting through a triangle create similar triangles.

Answer: (D) $\frac{2}{3}$

CIRCLES

1. How many *degrees* are in a circle?   <u>360</u>

2. What is the formula for *area* of a circle?   <u>πr^2</u>

3. What is the formula for *circumference* of a circle?   <u>$2\pi r$ or πd</u>

4. A line segment from the center of the circle to any point on the circle is called what?   <u>***A radius***</u>

5. What is a line segment drawn from one side of the circle to another side of the circle called?   <u>***A chord***</u>

6. What is the longest chord in a circle?   <u>***The diameter***</u>

7. What makes the diameter special?   <u>It crosses through the center of the circle.</u>

8. What is half of a circle called?   <u>***A semi-circle***</u>

9. What is a line that intersects a circle at exactly one point called?   <u>***A tangent***</u>

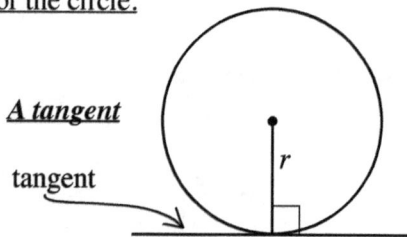

10. What is formed when a radius and a tangent intersect?   <u>A right angle</u>

One of ETS's favorite ways to test your knowledge of circles is to ask for *arc length* and *sector area*.

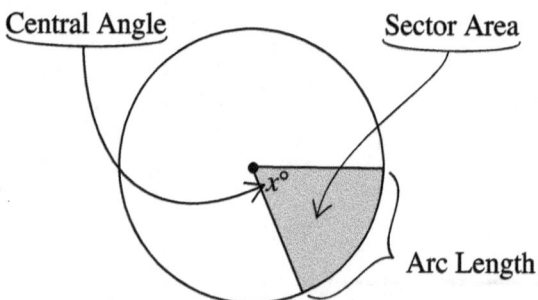

The arc is like the crust of a pizza, and the sector area is all the good stuff in a slice of the pizza.

Arc length and sector area are always relative to the ratio of the central angle to 360°. Memorize this proportion:

$$\frac{\text{Sector Area}}{\pi r^2} = \frac{\text{Central Angle }(x)}{360} = \frac{\text{Arc Length}}{2\pi r}$$

152

Let's say ETS gives us a circle with a central angle of 60° and asks, "What fractional part of the circle is arc $\overset{\frown}{AB}$?"

Knowing that arc length is always in relation to the central angle over 360, just put $\frac{60}{360}$ and reduce. Since the central angle is $\frac{1}{6}$ of the circle, $\overset{\frown}{AB}$ is $\frac{1}{6}$ of the total circumference of the circle, and the sector area would also be $\frac{1}{6}$ of the circle's total area.

$$\frac{\text{Central Angle}}{360} = \frac{60}{360} = \frac{6}{36} = \frac{1}{6}$$

When asked for a specific measurement for arc length and sector area, the proportions come into play. Let's say ETS gives us a circle with a central angle of 40° and a radius of 4, and asks for the arc length, a.

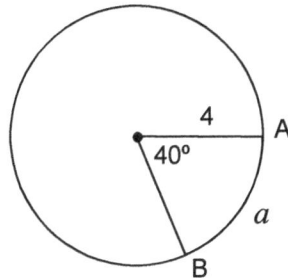

$$\frac{\text{Arc Length}}{2\pi r} = \frac{\text{Central Angle}}{360}$$
$$\frac{a}{2\pi(4)} = \frac{40}{360}$$
$$\frac{a}{8\pi} \times \frac{1}{9}$$
$$9a = 8\pi$$
$$\frac{9a}{9} = \frac{8\pi}{9}$$
$$a = \frac{8\pi}{9}$$

Let's try another one. Say ETS gives us a circle with central angle 90° and a sector area of 16π and asks us to find the radius.

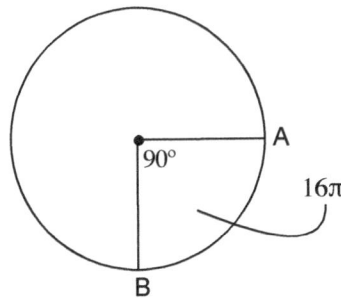

$$\frac{\text{Sector Area}}{\pi r^2} = \frac{\text{Central Angle}}{360}$$
$$\frac{16\pi}{\pi r^2} = \frac{90}{360}$$
$$\frac{16}{r^2} \times \frac{1}{4}$$
$$r^2 = 64$$
$$\sqrt{r^2} = \sqrt{64}$$
$$r = 8$$

Let's work an example:

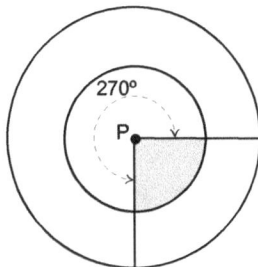

13. Point P is the center of both circles in the figure shown above. If the area of the large circle is 64 and the radius of the large circle is twice the radius of the small circle, what is the value of the shaded area?

(A) 2
(B) 4
(C) 8
(D) 12
(E) 16

153

Explanation:

Step 1:
large circle:
$\pi r^2 = 64$
Set the formula for area equal to the given area to solve for radius:

$$\frac{\pi r^2}{\pi} = \frac{64}{\pi}$$

$$\frac{\cancel{\pi} r^2}{\cancel{\pi}} = \frac{64}{\pi}$$

$$r^2 = \frac{64}{\pi}$$

Square root both sides:

$$\sqrt{r^2} = \frac{\sqrt{64}}{\sqrt{\pi}}$$

$$r = \frac{8}{\sqrt{\pi}}$$

Step 2:
small circle:
radius of small circle is half the radius of the big circle:

$$r = \frac{\frac{8}{\sqrt{\pi}}}{2}$$

$$r = \frac{8}{\sqrt{\pi}} \div \frac{2}{1}$$

Switch the operator:

$$r = \frac{8}{\sqrt{\pi}} \times \frac{1}{2}$$

Cross-cancel:

$$r = \frac{\overset{4}{\cancel{8}}}{\sqrt{\pi}} \times \frac{1}{\cancel{2}_1}$$

$$r = \frac{4}{\sqrt{\pi}} \times \frac{1}{1}$$

$$r = \frac{4}{\sqrt{\pi}}$$

Step 3:
The central angle of the shaded area is $360 - 270$, so central angle = $90°$. Use your proportion:

$$\frac{\text{central } \angle}{360°} = \frac{\text{area of shaded sector}}{\pi r^2}$$

$$\frac{90}{360} = \frac{x}{\pi \left(\frac{4}{\sqrt{\pi}}\right)^2} \quad \text{(sector area}$$

Reduce the left side and distribute the exponent on the right:

$$\frac{1}{4} = \frac{x}{\pi \left(\frac{4}{\sqrt{\pi}} \cdot \frac{4}{\sqrt{\pi}}\right)}$$

The square roots cancel out and we are just left with π:

$$\frac{1}{4} = \frac{x}{\pi \left(\frac{16}{\pi}\right)}$$

$$\frac{1}{4} = \frac{x}{\pi \left(\frac{16}{\cancel{\pi}}\right)}$$

$$\frac{1}{4} \diagdown \frac{x}{16}$$

Cross-multiply:
$$4x = 16$$

$$\frac{4x}{4} = \frac{16}{4}$$

$$x = 4$$

Answer: (B) 4

QUADRILATERALS

A quadrilateral is any four-sided figure. Squares, rectangles, parallelograms, trapezoids, and rhombuses are quadrilaterals. Even funky four-sided figures that don't have names are quadrilaterals.

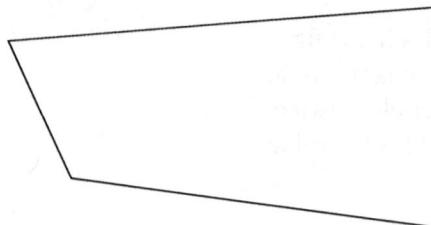

How many *degrees* are in a quadrilateral? <u>360</u>

ETS's favorite quadrilaterals are rectangles, squares, and parallelograms.

Rectangles

Each interior angle of a rectangle is equal to 90 degrees and opposite sides of a rectangle are equal.

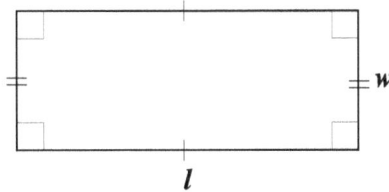

Area = lw
Perimeter = $2(l + w)$

Remember: **Perimeter** is the distance around an object and **Area** is the measurement of the inside portion of the figure.

Let's say ETS gives us a rectangle with a diagonal of length 10 that splits the rectangle into two 30-60-90 triangles. What is the area of the rectangle?

Explanation:
Step 1:
Draw in the diagonal and label what we know. Divide the hypotenuse by 2 to find the short side (the width of the rectangle):

long side $= 10$

$\frac{10}{2} = 5$

short side $= 5$

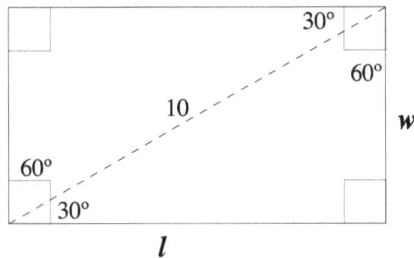

Step 2:
Now multiply by $\sqrt{3}$ to find the medium side (the length of the rectangle), which is $5\sqrt{3}$.

$5 \times \sqrt{3} = 5\sqrt{3}$
medium side $= 5\sqrt{3}$

Step 3:
Area of rectangle = lw
$A = 5 \times 5\sqrt{3}$
$A = 25\sqrt{3}$

Step 4:
Now that we've found the area of the rectangle, what is its perimeter?
Add all of the sides together:

$2(5 + 5\sqrt{3}) = 10 + 10\sqrt{3}$

perimeter $= 10 + 10\sqrt{3}$

Squares

A square is a special type of rectangle in which all sides are equal (all angles still equal 90°).

Area = s^2
Perimeter = $4s$

Remember: if you split a square along its diagonal it always forms two 45-45-90 triangles:

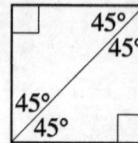

Parallelograms

A parallelogram has two pairs of parallel sides. Opposite angles are equal and consecutive angles add up to 180°.

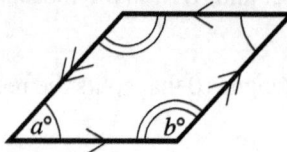

$a + b = 180$

You must remember to drop down the height to solve for the area. Do not plug in the slanted side as your height.

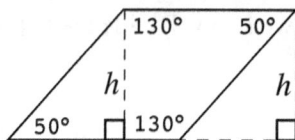

Area = bh

Perimeter = 2(*base* + *slanted side*)

If ETS tells you that the slant height of a parallelogram is 5 and gives you no way of solving for the actual height, then all you know is that the actual height of the parallelogram is less than the slant height of 5.

What is the area of the parallelogram?

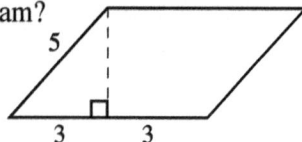

Notice the special right triangle within the parallelogram. We have a 3–4–5 triangle. So, $h = 4$ and $b = 6$. Now solve for the area:

Area = bh
= (4)(6)
Area = 24

Let's try it out:

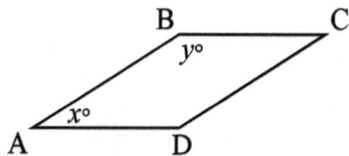

5. In the figure, ABCD is a rhombus. What is the value of $\frac{1}{2}(x + y)$?

    (A) 90
    (B) 120
    (C) 180
    (D) 270
    (E) 360

    Explanation:
    A rhombus is a parallelogram and a rule for parallelograms is that adjacent angles add up to 180°. So, $x + y = 180$

    $\frac{1}{2}(180) = 90$

    Answer: (A) 90

156

OTHER POLYGONS

ETS often gives problems that deal with figures with 5 or more sides. The problems require you to be able to find the sum of the degree measures of a polygon's interior angles. You can either use the formula: $(n-2)\,180$ where n stands for the number of sides, or you can pick a vertex and draw triangles to each of the other vertices. Count the number of triangles formed and then multiply by $180°$.

Let's see how both options work with an 8-sided figure:

Option 1

$(n-2)\,180$
$n = 8$
$(8-2)\,180 =$
$(6)180 = 1080$

Option 2

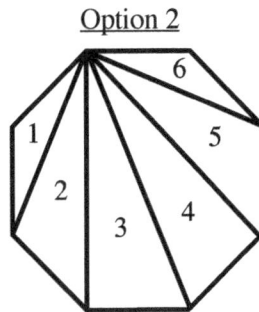

$(6)180 = 1080$

So, the sum of all angles in an octagon is $1080°$.

Let's try another polygon problem:

10. A nine-sided polygon has 8 equal angles. If the measure of one of the angles in the polygon is 100, what is the degree measure of each of the other 8 angles? (Disregard the degree symbol when gridding in your answer.)

Explanation:
$(n-2)180 =$
$(9-2)180 =$
$7 \cdot 180 = 1260$
Subtract the total degrees of the polygon from the degree measure of the given angle:
$1260 - 100 = 1160$
Divide by 8 to find each of the other degree measures:
$\dfrac{1160}{8} = 145$
Answer: 145

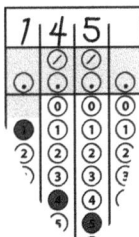

Note: A regular polygon has equal angles and equal side lengths. Each angle in a regular n-sided polygon equals the sum of the angles divided by the number of angles:

each angle $= \dfrac{(n-2)180}{n}$

FUNKY FIGURES

ETS often gives us funky figures and asks us to solve for the perimeter or area. The trick is to break the figure up into shapes you know.

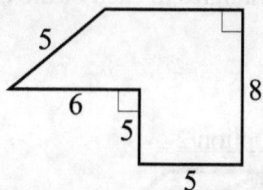

Note: Figure not drawn to scale

9. What is the area of the six-sided figure above?

(A) 26
(B) 52
(C) 58
(D) 134
(E) 140

Explanation:
Divide the funky shape into familiar shapes; find the areas of each of the familiar shapes and add them together. The key to this problem is recognizing the 3–4–5 special triangle.

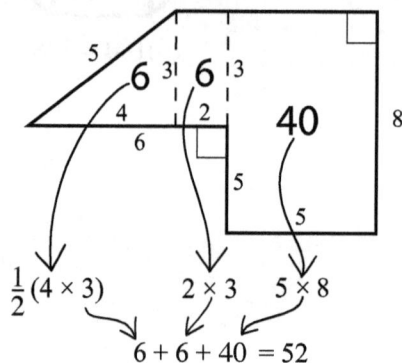

$$\frac{1}{2}(4 \times 3) \qquad 2 \times 3 \qquad 5 \times 8$$

$$6 + 6 + 40 = 52$$

Answer: (B) 52

OVERLAPPING FIGURES

And then there is the oh-so-common overlapping figure problem. This is where you have to get really good at putting the pieces of the puzzle together. You already know the individual rules and formulas so let's have at it.

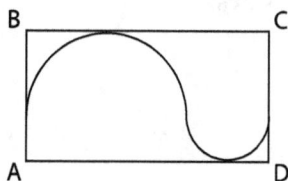

6. The perimeter of rectangle ABCD above is 18 and the length of $\overline{CD}$ is 3. What is the total length of the curve?

(A) 2π
(B) 3π
(C) 6π
(D) 9π
(E) 18π

158

Explanation:

We are given a width of 3 and a perimeter of 18, so the length must be 6:

$2(width) + 2(length) = perimeter$
$2(3) + 2l = 18$
$6 + 2l = 18$
$6 + 2l - 6 = 18 - 6$
$2l = 12$
$\frac{2l}{2} = \frac{12}{2}$
$l = 6$

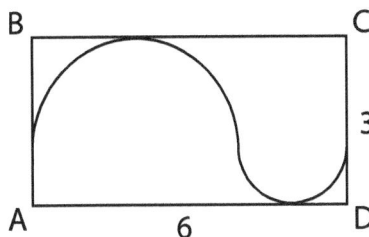

From here it's a Sneaky Plug In. Let's break up that 6 into two diameters, one for the large circle and one for the small circle:

d of the big semi-circle = 4
d of the small semi-circle = 2

Big semi-circle: $c = 2\pi r = 2\pi(2) = 4\pi$
Since it's a semi-circle, we have to divide by 2: $\frac{4\pi}{2} = 2\pi$

Small semi-circle: $c = 2\pi r = 2\pi(1) = 2\pi$
Since it's a semi-circle, we have to divide by 2: $\frac{2\pi}{2} = \pi$

Length of the curve = circumference of the big semi-circle + circumference of the small semi-circle

Length of the curve = $2\pi + \pi = 3\pi$

Answer: (B) 3π

Note: We could have broken up the 6 into 5 and 1, or even 3 and 3, and we would still arrive at the same answer, 3π.

SHADED AREAS

A variation of the overlapping figure problem is the shaded area problem.

The key to shaded area problems is to identify the two or three given shapes. Often it comes down to simply subtracting the area of the smaller shape from the area of the larger shape to find the shaded area. Also, you should always ask: *what do these two shapes have in common?*

For instance, let's say a square is inscribed in a circle and the shaded area is the part of the circle not taken up by the square. ETS tells us that a side of the square equals 4. What is the shaded area?

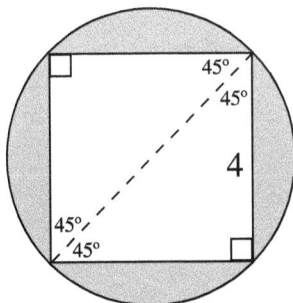

area of square $= s^2$
area of square $= 4^2$
$\qquad = 16$
diagonal of square $= 4 \times \sqrt{2}$
$\qquad = 4\sqrt{2}$

Notice that the diagonal of the square is the diameter of the circle.

radius of circle $= 4\sqrt{2} \div 2 = \frac{{}^2\cancel{4}\sqrt{2}}{\cancel{2}_1} = 2\sqrt{2}$
area of circle $= \pi r^2$
area of circle $= \pi(2\sqrt{2})^2$
$\qquad = \pi(2\sqrt{2} \times 2\sqrt{2}) = \pi(4 \times 2) = 8\pi$

shaded area = area of circle − area of square
shaded area $= 8\pi - 16$

Try one on your own:

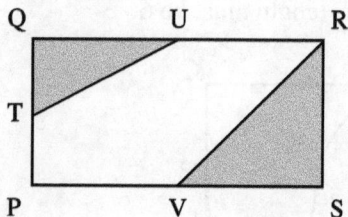

Q   U   R

T

P   V   S

15. In rectangle PQRS above, T is the midpoint of side PQ, U is the midpoint of side QR, and V is the midpoint of side PS. What fraction of the area of the rectangle is unshaded?

(A) $\frac{2}{3}$

(B) $\frac{4}{6}$

(C) $\frac{3}{5}$

(D) $\frac{3}{8}$

(E) $\frac{5}{8}$

Explanation:
This is a Sneaky Plug In! Start by plugging in for the sides of the rectangle: Let the length = 8 and the width = 4. (Any numbers will do, as long as you obey the restrictions.)

Step 1:
find total area
lw = area
$8 \times 4 = 32$

Step 2:
Find the area of $\triangle$QTU
QP = 4
So, QT = 2
QR = 8
So, QU = 4
$\frac{1}{2}(4 \times 2) =$
$\frac{1}{2}(8) = 4$

Step 3:
Find the area of $\triangle$RSV
PS = 8
So, VS = 4
RS = 4
$\frac{1}{2}(4 \times 4) =$
$\frac{1}{2}(16) = 8$

Step 4:
Find the unshaded area
Unshaded area = Area of PQRS − (Area of $\triangle$QTU + Area of $\triangle$RSV)
Unshaded area = 32 − (4 + 8)
$\qquad\qquad\quad$ = 32 − 12
$\qquad\qquad\quad$ = 20

$\frac{\text{Unshaded area}}{\text{Area of PQRS}} = \frac{20}{32} = \frac{5}{8}$

Answer: (E) $\frac{5}{8}$

3D FIGURES

Here are the 3D figures you need to know: ***Rectangular Solids, Cubes, Cylinders, Spheres***

Rectangular Solids and Cubes

1. What is the formula for ***volume of a rectangular solid***? $\underline{l \times w \times h}$

2. What is the formula for ***volume of a cube***? $\underline{s^3}$

3. What is ***surface area***?   <u>The sum of the areas of all the outside faces of a 3D solid</u>

4. What is the formula for ***surface area of a rectangular solid***? $\underline{2lw + 2lh + 2wh}$

5. What is the formula for ***surface area of a cube***? $\underline{6s^2}$

160

Surface Area formulas are not in the reference box at the beginning of each section so be sure you know them.

Let's try a rectangular solid problem:

12. The top and bottom of a rectangular solid are painted blue, and the remaining faces are painted white. The total area of the blue faces is 36 square inches. If the height of the rectangular solid is 4, what is the volume of the rectangular solid in cubic inches?

(A) 54
(B) 72
(C) 98
(D) 108
(E) 216

Explanation:

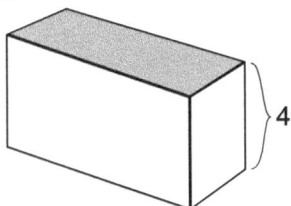

Volume = $l \times w \times h$

Total area of blue faces = 36
But there are two blue faces, so the area of each blue face is $\frac{36}{2} = 18$
The length and width can be any factor pair of 18, such as 6 and 3.

$6 \times 3 \times 4 = 72$

Answer: (B) 72

Notice that we don't have to choose 6 and 3 as the length and width; any factor pair of 18 works:

2 and 9 are also factors of 18:
$2 \times 9 \times 4 = 72$ ✓

Even 18 and 1 work:
$18 \times 1 \times 4 = 72$ ✓

It's important to note that the greatest length in a rectangular solid or cube is the ***long diagonal***. Memorize the formula for the long diagonal:

Super Pythagorean Theorem: $d^2 = a^2 + b^2 + c^2$, where a is the length, b is the width, and c is the height.

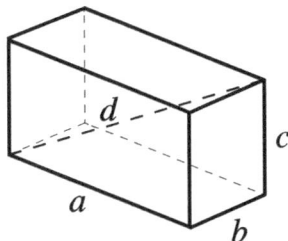

CYLINDERS

What is the formula for ***volume of a cylinder***? $\pi r^2 h$

Let's try a cylinder problem:

12. Bucket A and Bucket B are right circular cylinders. The radius of Bucket B is ⅓ of the radius of Bucket A, and the height of Bucket B is ¼ of the height of Bucket A. If the radius of Bucket A is 9, the height of Bucket A is 8, and 2 vases of water completely fill up Bucket B, how many vases of water fill up Bucket A?

(A) 29
(B) 54
(C) 72
(D) 15
(E) 96

Explanation:
Volume of Bucket A:
$\pi r^2 h$ = volume
$\pi 9^2 \cdot 8 =$
$\pi 81 \cdot 8 = 648\pi$

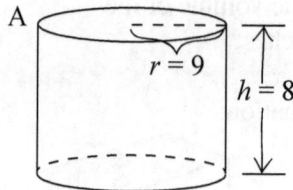

Dimensions of Bucket B:
⅓(9) = 3 = radius
¼(8) = 2 = height

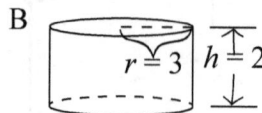

Volume of Bucket B:
$\pi r^2 h$ = volume
$\pi 3^2 \cdot 2 =$
$\pi 9 \cdot 2 = 18\pi$

Since 2 vases of water fill up Bucket B then 2 vases of water is equal to 18π.

$\frac{648\pi}{18\pi} = 36$, and since there are two vases in 18π, multiply by 2: $36 \times 2 = 72$.

Answer: (C) 72

Alternatively, we can set up a proportion:
$$\frac{2 \text{ vases}}{18\pi} \,\,\times\,\, \frac{x \text{ vases}}{648\pi}$$
$18\pi x = 1296\pi$
$\frac{18\pi x}{18\pi} = \frac{1296\pi}{18\pi}$
$x = 72$

SPHERES

Spheres have been making guest-star appearances on the SAT quite a bit. To the best of my knowledge, no sphere formulas have been needed thus far, but here they are just in case.

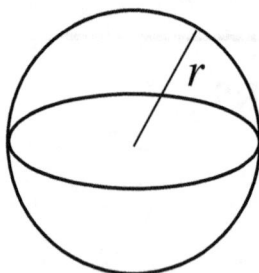

Surface area of a sphere = $4\pi r^2$

Volume of a sphere = $\frac{4}{3}\pi r^2$

Let's try a sphere problem:

5. S is a point on a sphere with radius 11 and T is a point on a sphere with radius 6. If the two spheres are tangent to each other, what is the greatest possible length of $\overline{ST}$?

(A) 11

(B) 16   Explanation:

(C) 17   Draw the two spheres. Since we want the greatest length of $\overline{ST}$, we want S and T to be as

(D) 22   far away from each other as possible. If we connect the two diameters in a straight line and

(E) 34   put S and T at opposite ends:

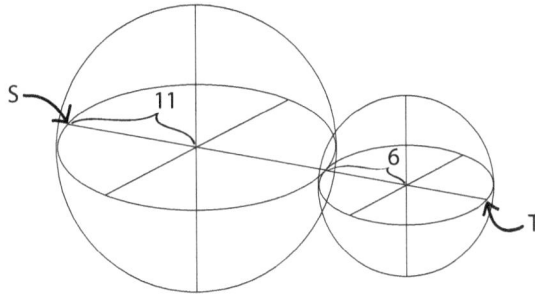

$11 + 11 + 6 + 6 = 34$

Answer: (E) 34

COORDINATE GEOMETRY

Another big portion of SAT Math is devoted to Coordinate Geometry.

A point is denoted as the ordered pair (x, y) where the x coordinate tells you how far to the right or left of the origin to move on the x-axis and the y coordinate tells you how far up and down from the origin to move on the y-axis.

Let's plot some coordinate points:

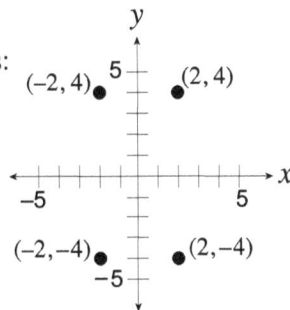

You should also be familiar with the 4 quadrants of the coordinate plane:

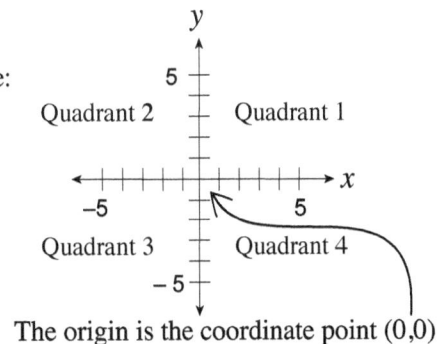

Points in Quadrant 1 have positive x and positive y coordinates.
Points in Quadrant 2 have negative x but positive y coordinates.
Points in Quadrant 3 have negative x and negative y coordinates.
Points in Quadrant 4 have positive x but negative y coordinates.

The origin is the coordinate point (0,0)

163

Whenever ETS wants you to find the distance between two points, don't worry about using the distance formula. Just draw in a right triangle. The distance is always the hypotenuse and it's easy to use the Pythagorean Theorem (which is the basis for the distance formula anyway!).

Let's say ETS asks, "What is the distance between (2,6) and (4,1)?"

Plot the 2 points and draw in a right triangle:

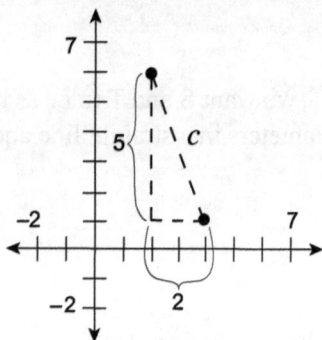

The base of the triangle is 2, and the height is 5. We can plug these into the Pythagorean Theorem and solve for the hypotenuse, which is the distance between the two points:

$$5^2 + 2^2 = c^2$$
$$25 + 4 = c^2$$
$$29 = c^2$$
$$\sqrt{29} = \sqrt{c^2}$$
$$c = \sqrt{29}$$

---

9. In the *xy*-coordinate plane the distance from point S to point $(-2, 3)$ is 7. If the x-coordinate of S is -2, which of the following could be the y-coordinate of S?

    (A) 6
    (B) 4
    (C) 0
    (D) 2
    (E) -4

Explanation:
Draw the graph:

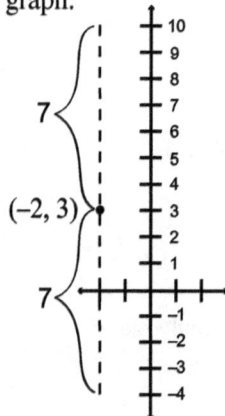

The two possibilities are 10 and -4, but only -4 is an answer choice.

Answer: (E) -4

164

To find the **midpoint** of a line segment use this formula: $\left(\dfrac{x_1 + x_2}{2}, \dfrac{y_1 + y_2}{2}\right)$

Basically, you find the average of the x coordinates and of the y coordinates of the endpoints.

The midpoint of $(2, -3)$ and $(4, t)$ is $(3, 4)$. What is the value of t?

Apply the formula $\left(\dfrac{y_1 + y_2}{2} = midpoint\right)$: $\dfrac{-3+t}{2} = 4$

$$\dfrac{-3+t}{2} \times \dfrac{4}{1}$$
$$-3 + t = 8$$
$$-3 + t + 3 = 8 + 3$$
$$t = 11$$

SLOPE

Remember **slope** as **rise over run** – rise meaning how many points up or down, and run meaning how many points right or left.

The **formula for finding slope** of a line is $\dfrac{y_2 - y_1}{x_2 - x_1}$

What is the slope of the line that contains the coordinate points $(2, 3)$ and $(-5, 5)$?

Just plug the points into the formula: $\dfrac{5-3}{-5-2} = \dfrac{2}{-7} = -\dfrac{2}{7}$

Here are the visual representations of positive slope, negative slope, zero slope, and undefined slope. Knowing these will help big time with process of elimination on graphing problems.

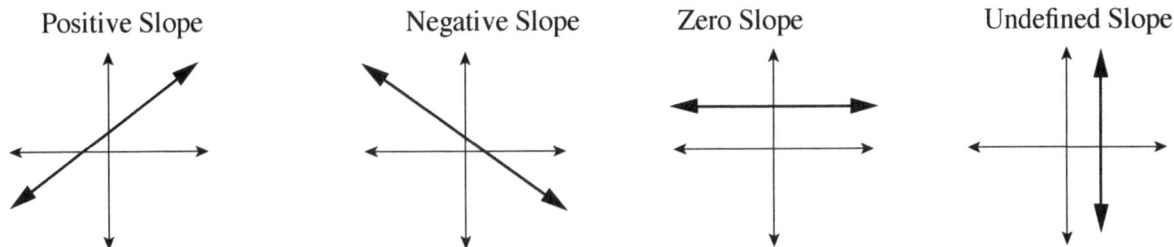

Positive Slope	Negative Slope	Zero Slope	Undefined Slope

1. What is the relationship between the **slopes of two parallel lines**?   <u>They are equal</u>

2. What is the relationship between the **slopes of two perpendicular lines**?   <u>They are negative reciprocals of each other</u>

Understanding this information, let's see how to apply it to the following problem:

14. Line r lies in the xy-coordinate plane and contains points $(3, 5)$ and $(-2, t)$. If line r is perpendicular to line s, and the slope of line s is -5, what is the value of t?

    (A) -4
    (B) -2
    (C) 2
    (D) 4
    (E) 5

Explanation:
Perpendicular lines have negative reciprocal slopes, so the slope of line r is $\frac{1}{5}$.

Option 1:
Use the slope formula:

$$\frac{y_2 - y_1}{x_2 - x_1}$$

$$\frac{t - 5}{-2 - 3} = \frac{1}{5}$$

$$\frac{t - 5}{-5} = \frac{1}{5}$$

Option 2:
From here, we can also solve by plugging in for t. For example, plugging in (D) 4:

$$5(t - 5) = -5$$

$$\frac{4 - 5}{-2 - 3} = \frac{1}{5}$$

$$\frac{5(t - 5)}{5} = \frac{-5}{5}$$

$$\frac{-1}{-5} = \frac{1}{5}$$

$$\frac{\cancel{5}(t - 5)}{\cancel{5}} = \frac{-\cancel{5}}{\cancel{5}}$$

$$\frac{1}{5} = \frac{1}{5} \checkmark$$

$$t - 5 = -1$$

$$t - 5 + 5 = -1 + 5$$

$$t = 4$$

Answer: (D) 4

But graphing lines is more than just understanding slope.

3. What is the equation of a line in *slope-intercept form*? $y = mx + b$

4. What does m equal?   slope

5. What does b equal? y intercept

The *y intercept* means the point where the line crosses the y-axis (the value of y when $x = 0$) and the *x intercept* means the point where the line crosses the x-axis (the value of x when $y = 0$).

Let's say ETS gives you the equation $y = 2x + 4$. Graph the line using the intercepts:

Step 1:
Find the y intercept.

The y intercept is 4 or $(0, 4)$

Step 2:
To find the x intercept plug in 0 for y and solve for x.

So $0 = 2x + 4$

$$0 - 4 = 2x + 4 - 4$$

$$-4 = 2x$$

$$\frac{-4}{2} = \frac{2x}{2}$$

$$x = -2$$

x intercept is -2 or $(-2, 0)$

Step 4:
Graph the line by plotting your x and y intercepts and connecting the dots.

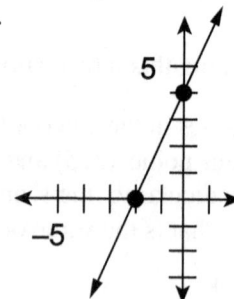

Sometimes the equation of a line will be given in Standard Form: $Ax + By + C = 0$. In these cases, your first step should be to convert to slope-intercept form.

If you see $2x + 3y = 10$, convert to slope intercept-form and solve for y.

$$2x + 3y = 10$$

$$2x + 3y - 2x = -2x + 10$$

$$3y = -2x + 10$$

$$\frac{3y}{3} = \frac{-2x + 10}{3}$$

$$y = -\frac{2}{3}x + \frac{10}{3}$$

Once you plot your y-intercept, use the slope to plot a second point (down 2, to the right 3).

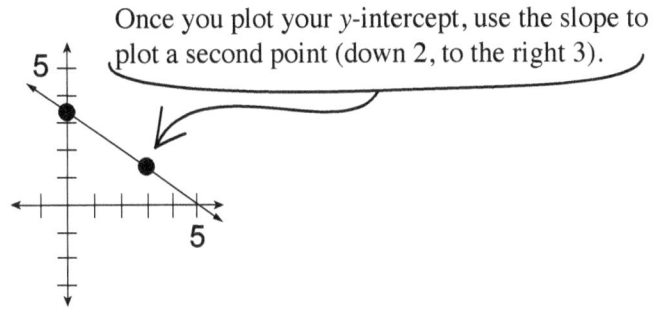

Let's try more problems:

10. In the xy plane line $5x - 6y = b$ passes through point $(4, -2)$. What is the value of b?

Explanation:
Plug in the given values for x and y: $x = 4$ and $y = -2$
$$5x - 6y = b$$
$$5(4) - 6(-2) = b$$
$$20 + 12 = b$$
$$32 = b$$

Answer: 32

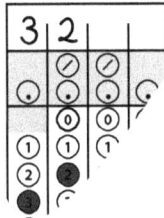

20. The equation of line m is $2x + 3y = 6$ and line n is perpendicular to line m. If line n contains the coordinate points $(0, 3)$ and $(2, r)$, what is the value of r?

(A) -6

(B) $-\frac{5}{3}$

(C) 0

(D) $\frac{5}{3}$

(E) 6

Explanation:
First put the equation for line m into $y = mx + b$ form:

$$2x + 3y = 6$$

$$2x + 3y - 2x = -2x + 6$$

$$3y = -2x + 6$$

$$\frac{3y}{3} = \frac{-2x + 6}{3}$$

$$y = -\frac{2}{3}x + 2$$

167

Perpendicular lines have negative reciprocal slopes so the slope of line n is 3/2. Plug this value into the slope formula and solve for r:

$$\frac{y_2 - y_1}{x_2 - x_1} = slope$$

$$\frac{r-3}{2-0} = \frac{3}{2}$$

$$\frac{r-3}{2} \diagdown \frac{3}{2}$$

$$2(r-3) = 6$$

$$2r - 6 = 6$$

$$2r - 6 + 6 = 6 + 6$$

$$2r = 12$$

$$\frac{2r}{2} = \frac{12}{2}$$

$$r = 6$$

Answer: (E) 6

We could also finish solving with the A.C.T to find which answer gives us the correct slope of 3/2:

(C) 0

$$\frac{0-3}{2-0} = \frac{-3}{2} = -\frac{3}{2} \ \times$$

(D) $\frac{5}{3}$

$$\frac{\frac{5}{3}-3}{2-0} = \frac{-1\frac{1}{3}}{2} = -\frac{4}{6} \ \times$$

(E) 6

$$\frac{6-3}{2-0} = \frac{3}{2} \ \checkmark$$

$$dx + 5y = -10$$

15. The equation of a line in the xy-plane is shown above. If d is a constant and the slope of the line is -7, what is the value of d?

(A) 22
(B) 37
(C) 41
(D) 24
(E) 35

Explanation:

Step 1:

Transform the equation into $y = mx + b$ form:

$$dx + 5y = -10$$

$$dx + 5y - dx = -dx - 10$$

$$5y = -dx - 10$$

$$\frac{5y}{5} = \frac{-dx - 10}{5}$$

$$y = \frac{-dx - 10}{5}$$

$$y = -\frac{dx}{5} - 2$$

Answer: (E) 35

Step 2:

$-d/5$ is the slope and the slope is equal to -7. So, set $-d/5$ equal to -7 and solve for d:

$$\frac{-d}{5} = -7$$

$$\frac{-d}{5} \diagdown \frac{-7}{1}$$

$$-d = -35$$

$$\frac{-d}{-1} = \frac{-35}{-1}$$

$$d = 35$$

REFLECTIONS

Reflection problems on the SAT boil down to finding a figure's mirror image when it is reflected across a line.

Here are some rules to know:

If a line is **reflected across the x-axis**, both the slope and the y-intercept are multiplied by -1.

The reflection of point (x, y) across the x-axis is the point $(x, -y)$.

For example: If $y = 2x - \frac{2}{3}$ is reflected across the x-axis, the new equation of the reflected line would be $y = -2x + \frac{2}{3}$

Here's what that reflection would look like:

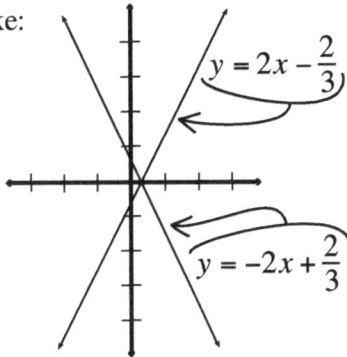

Here is a visual example of a triangle reflected across the x-axis:

If ever in doubt, simply fold your paper along the line of the reflection (in this example, the x-axis) to see where your new figure will be located.

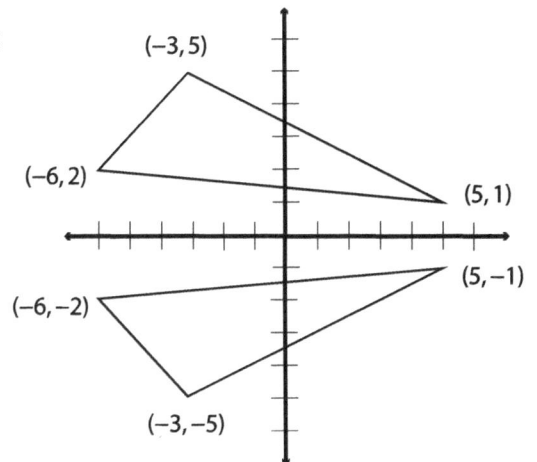

Here is a visual example of a triangle **reflected across the y-axis**:

If a point is reflected across the y-axis, the y-coordinate remains the same, but the x-coordinate is transformed into its opposite. The slopes of lines reflected about the y-axis are negatives of each other.

The reflection of the point (x, y) across the y-axis is the point $(-x, y)$.

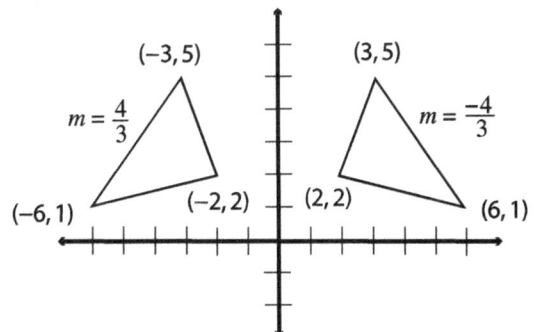

Another Reflection concept that has been tested recently on the SAT is reflecting over the line $y = x$:

Here is a visual of a triangle reflected over the line $y = x$:

When you **reflect a point across the line $y = x$**, the x-coordinate and the y-coordinate change places.

The reflection of the point (x, y) across the line $y = x$ is the point (y, x).

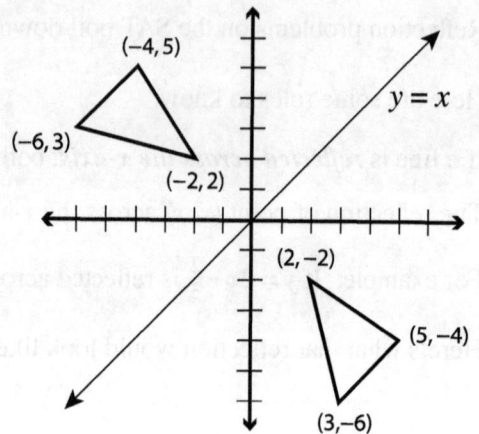

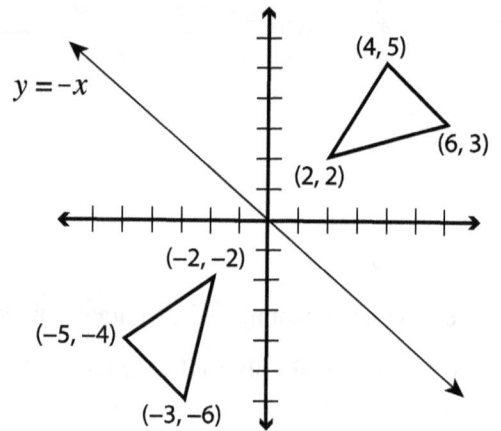

Here is a visual of a triangle reflected over the line $y = -x$:

When you **reflect a point across the line $y = -x$**, the x-coordinate and the y-coordinate change places and are negated (the signs are changed).

The reflection of the point (x, y) across the line $y = -x$ is the point $(-y, -x)$.

Let's do some SAT reflection problems:

13. Three vertices of a triangle that lies in the xy-coordinate plane are $(0, 0), (4, 0)$, and $(3, 4)$. If the triangle is reflected about the line $y = -x$, what is one vertex of the reflection?

(A) $(-4, 3)$
(B) $(-3, -4)$
(C) $(0, 3)$
(D) $(0, 4)$
(E) $(-4, -3)$

Explanation:
The reflection of the point (x, y) across the line $y = -x$ is the point $(-y, -x)$ so $(3, 4)$ becomes $(-4, -3)$.

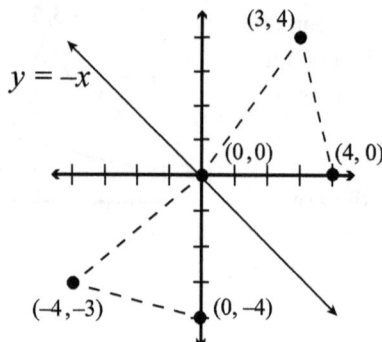

Answer: (E) $(-4, -3)$

170

14. In the *xy*-coordinate plane, the equation of line *m* is $y = 3x - 7$. If line *n* is a reflection of line *m* in the *y*-axis what is the equation of line *n*?

(A) $y = \dfrac{-1}{3x - 7}$

(B) $y = \dfrac{-1}{3x + 7}$

(C) $y = -3x - 7$

(D) $y = -3x + 7$

(E) $y = 3x + 7$

Explanation:
When a line is reflected across the *y*-axis the slope is a negative of the original slope and the *y*-intercept stays the same, so the slope should be −3 and the *y*-intercept should stay −7.

Answer: (C) $y = -3x - 7$

Let's do an 18-question geometry drill to test how well we can put together the concepts we've learned so far.

Geometry Drill

7. What is the maximum number of cube-shaped boxes with edges measuring 5 inches that can fit inside a rectangular box with interior measurements 20 inches by 15 inches by 5 inches?

   (A) 12
   (B) 24
   (C) 30
   (D) 45
   (E) 60

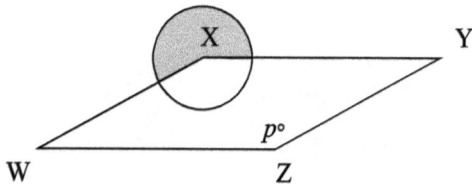

NOTE: Figure not drawn to scale

8. In the figure above, parallelogram WXYZ intersects the circle at its center point X. If $p = 130$, how many degrees constitute the shaded portion of the circle?

   (A) 180
   (B) 200
   (C) 230
   (D) 290
   (E) 310

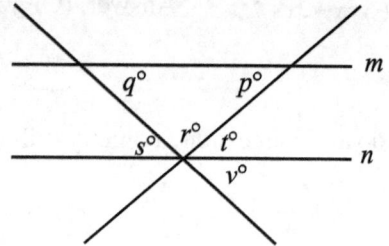

8. In the figure above, $m \| n$. Which of the following is equal to 180°?

   (A) $q + p + v$
   (B) $t + v + s$
   (C) $q + p + t$
   (D) $p + r + v$
   (E) $q + r + v$

9. How many points are a distance of 6 units from the origin in the xy-coordinate plane?

   (A) One
   (B) Two
   (C) Three
   (D) Four
   (E) More than four

$$-2 > -5x + 8$$

10. Which of the following represents all values of *x* that satisfy the inequality above?

(A)
 -2 -1 0 1 2

(B)
 -2 -1 0 1 2

(C)
 -2 -1 0 1 2

(D)
 -2 -1 0 1 2

(E)
 -2 -1 0 1 2

11. If point O lies in plane R, how many circles are there in plane R that have center O and an area of 16π inches?

(A) None
(B) One
(C) Two
(D) Four
(E) More than four

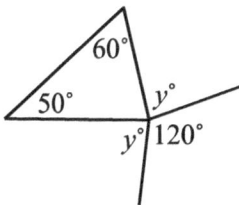

12. What is the value of *y* in the figure above?

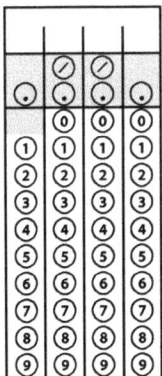

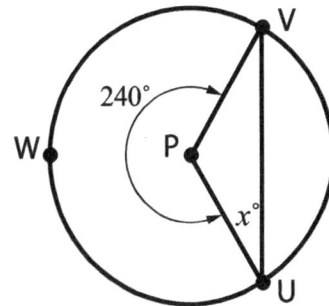

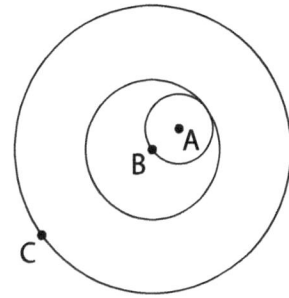

13. The centers of the three circles above lie on segment $\overline{AC}$. The center of the largest circle is point B, and the center of the smallest is point A. The middle circle and largest circle are concentric. If the middle circle bisects the radius of the largest circle, and the radius of the largest circle is 8, what is the area of the smallest circle?

(A) π
(B) 2π
(C) 3π
(D) 4π
(E) 5π

15. In circle P above, the measure of arc $\overset{\frown}{UWV}$ is $240°$. What is the value of *x*?

(A) 30
(B) 45
(C) 60
(D) 90
(E) 120

173

15. A can in the shape of a right circular cylinder has a height of 9 and a base with a diameter of 6. What is the distance from the center of the base to a point on the circumference of the top of the can?

(A) 8.48528
(B) 8.58629
(C) 9.48683
(D) 9.63542
(E) 10.81665

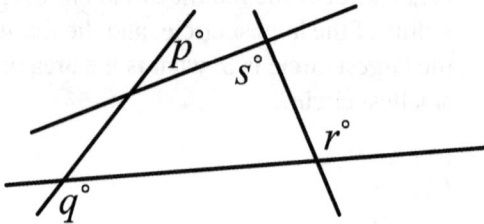

15. Four lines intersect as shown in the figure above. If $s = 100$, what is the sum of p, q, and r?

(A) 180
(B) 200
(C) 280
(D) 360
(E) It cannot be determined from the given information.

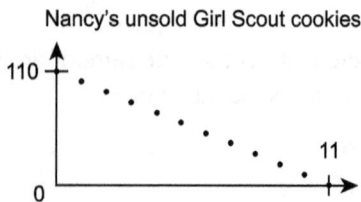

Nancy's unsold Girl Scout cookies

16. The number of Nancy's unsold Girl Scout cookies in an eleven day period is shown in the graph above. Which of the following equations best represents Nancy's unsold cookies?

(A) $y = 11x + 110$
(B) $y = 11x - 110$
(C) $y = 10x + 110$
(D) $y = 110 - 11x$
(E) $y = 110 - 10x$

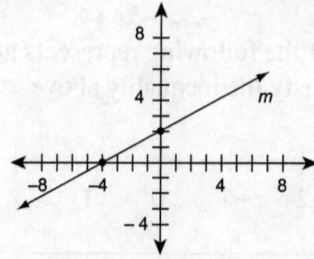

16. In the xy-plane, the equation of line m is $2y - x = 4$. Which of the following is an equation of a line that is perpendicular to line m ?

(A) $y = 2x + 6$
(B) $y = -2x + 6$
(C) $y = \frac{1}{2}x - 6$
(D) $y = \frac{1}{2}x + 4$
(E) $y = x + 4$

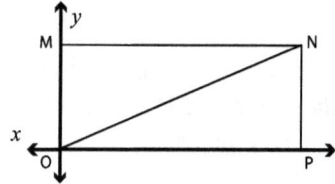

16. In the xy-plane above MNPO is a rectangle, $\triangle ONP$ has an area of 20, and the coordinates of P are $(10, 0)$. What is the slope of $\overline{ON}$?

(A) $\frac{3}{5}$
(B) 3
(C) $\frac{2}{5}$
(D) 5
(E) $\frac{2}{3}$

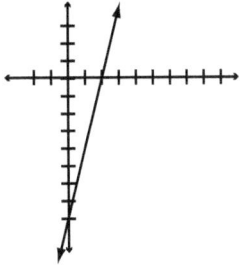

17. The equation of the line on the xy-coordinate graph is $y = 4x - 8$. Which of the following is the graph of $y = |4x - 8|$?

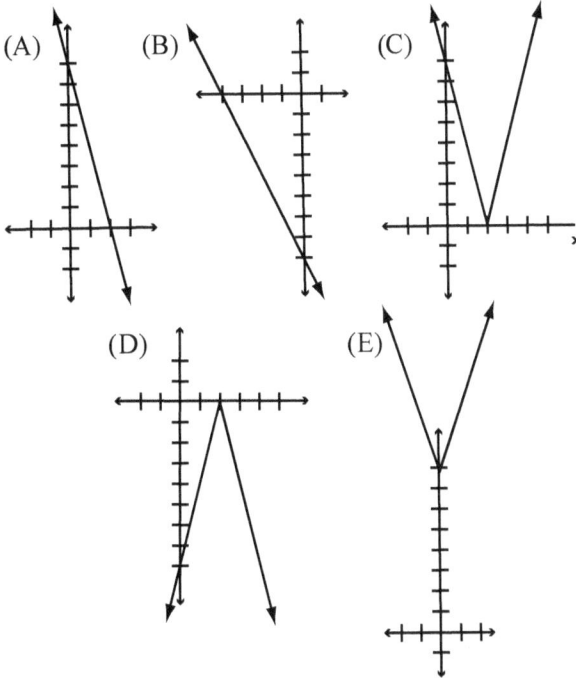

(A)

(B)

(C)

(D)

(E)

17. A trough in the shape of a rectangular solid has interior measurements of length 7 inches, width 3π inches, and height 4 inches. This trough is completely filled with water. All of the water is then poured into a container with an interior in the shape of a right circular cylinder with a radius of 4 inches. What must be the minimum inside height of the container?

(A) $\frac{4}{3}$

(B) $\frac{7}{3}$

(C) 2

(D) 4

(E) $\frac{21}{4}$

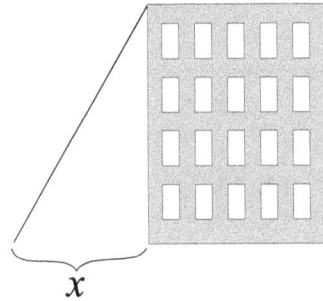

18. A ladder is extended from the roof of a building to the ground. The building is 1752.5 feet tall and the ladder has a slope of $\frac{5}{2}$. What is x, in feet?

(A) 350.5
(B) 701
(C) 876.25
(D) 1402
(E) 3505

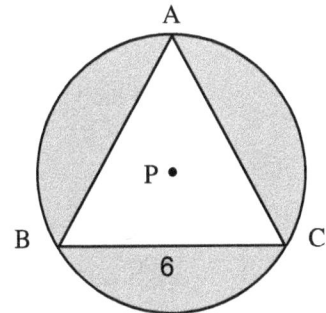

20. In the figure above, equilateral triangle ABC is inscribed in the circle with center P. What is the area of the shaded portion of the circle?

(A) $\frac{9\pi}{2} - 9\sqrt{3}$

(B) $\frac{9\pi}{2} - \frac{9\sqrt{3}}{2}$

(C) $12\pi - 9\sqrt{3}$

(D) $12\pi - \frac{9\sqrt{3}}{2}$

(E) $\frac{27\pi}{2} - \frac{9\sqrt{3}}{2}$

175

Answers and Explanations

7. What is the maximum number of cube-shaped boxes with edges measuring 5 inches that can fit inside a rectangular box with interior measurements 20 inches by 15 inches by 5 inches?

(A) 12

(B) 24

(C) 30

(D) 45

(E) 60

Explanation:

Find the volume of the rectangular solid:

$20 \times 15 \times 5 = 1500$

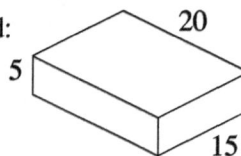

Find the volume of the cube:

$5 \times 5 \times 5 = 125$

Divide the volume of the solid by the volume of the cube(s):

$\frac{1500}{125} = 12$

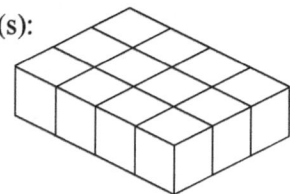

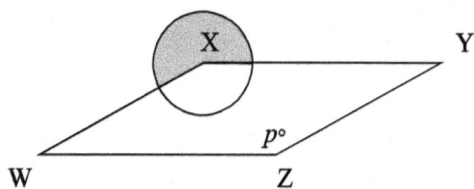

NOTE: Figure not drawn to scale

8. In the figure above, parallelogram WXYZ intersects the circle at its center point X. If $p = 130$, how many degrees constitute the shaded portion of the circle?

(A) 180

(B) 200

(C) 230

(D) 290

(E) 310

Explanation:

Opposite angles of a parallelogram are equal and there are 360° in a circle so:

$360 - 130 = 230$

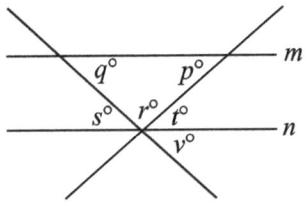

8. In the figure above, $m \parallel n$. Which of the following is equal to 180°?

(A) $q + p + v$
(B) $t + v + s$
(C) $q + p + t$
(D) $p + r + v$ ⟵ (circled)
(E) $q + r + v$

Explanation:
Sneaky Plug In! Plug In using your Big and Small angle rules for two parallel lines cut by a transversal.

Step 1:
Start by Plugging In $s = 40$
s and v are vertical angles, so $v = 40$
Plug In $r = 80$
$80 + 40 = 120$ and $s + r + t = 180$, so
$120 + t = 180$
$120 + t - 120 = 180 - 120$
$t = 60$

Step 2:
t and p are both alternate interior angles, so $p = 60$. There are 180° in a triangle
$q + r + p = 180$
$q + 80 + 60 = 180$
$q + 140 = 180$
$q + 140 - 140 = 180 - 140$
$q = 40$

Step 3:
So $s = 40°$, $v = 40°$, $r = 80°$, $t = 60°$, $q = 40°$, $p = 60°$

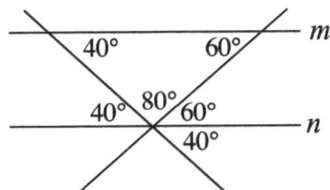

Now Plug In to your answer choices to find the sum that equals 180°

(A) $q + p + v$
    $40 + 60 + 40 = 140$ ✕

(B) $t + v + s$
    $60 + 40 + 40 = 140$ ✕

(C) $q + p + t$
    $40 + 60 + 60 = 160$ ✕

(D) $p + r + v$
    $60 + 80 + 40 = 180$ ✓

(E) $q + r + v$
    $40 + 80 + 40 = 160$ ✕

177

9. How many points are a distance of 6 units from the origin in the *xy*-coordinate plane?

(A) One
(B) Two
(C) Three
(D) Four
(E) More than four

Explanation:
A circle is an infinite set of points. If the radius is 6, all of the points in a circle are 6 units from the origin. Therefore, since the number of points is infinite, it is (E) More than four.

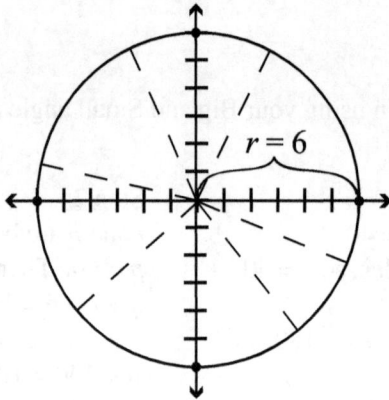

$r = 6$

$-2 > -5x + 8$

10. Which of the following represents all values of *x* that satisfy the inequality above?

(A)
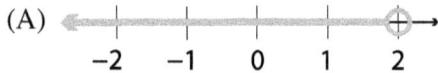
-2 -1 0 1 2

(B)
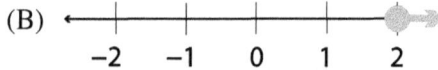
-2 -1 0 1 2

(C)
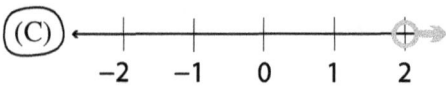
-2 -1 0 1 2

(D)
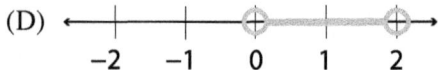
-2 -1 0 1 2

(E)
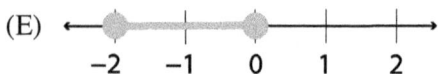
-2 -1 0 1 2

Explanation:
$-2 > -5x + 8$
$-2 - 8 > -5x + 8 - 8$
$-10 > -5x$
$\dfrac{-10}{-5} > \dfrac{-5x}{-5}$
$2 < x$

Remember: multiplying or dividing by a negative flips the inequality sign.

Note: filled in circles indicate $\leq$ or $\geq$, and open circles indicate $<$ or $>$.

178

11. If point O lies in plane R, how many circles are there in plane R that have center O and an area of 16π inches?

(A) None
(B) One
(C) Two
(D) Four
(E) More than four

Explanation:

Area = πr^2

$16\pi = \pi r^2$

$\dfrac{16\pi}{\pi} = \dfrac{\pi r^2}{\pi}$

$\dfrac{16\cancel{\pi}}{\cancel{\pi}} = \dfrac{\cancel{\pi} r^2}{\cancel{\pi}}$

$r^2 = 16$

$\sqrt{r^2} = \sqrt{16}$

$r = 4$

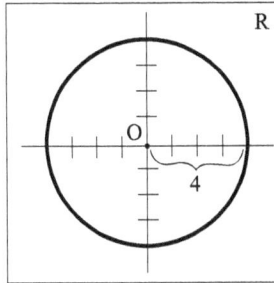

While we can draw many circles with a radius of 4, it is possible to draw only one circle with a radius of 4 and a center at point O.

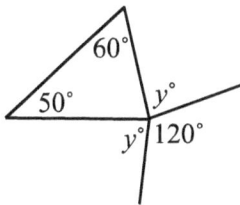

12. What is the value of y in the figure above?

Explanation:

Find the missing angle measure of the triangle:

$60 + 50 = 110$

$180 - 110 = 70$

Notice how $y + 70 + y + 120$ all form a circle:

$2y + 70 + 120 = 360$

$2y + 190 = 360$

$2y + 190 - 190 = 360 - 190$

$2y = 170$

$\dfrac{2y}{2} = \dfrac{170}{2}$

$y = 85$

Answer: 85

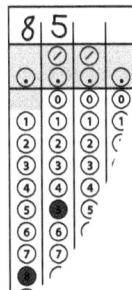

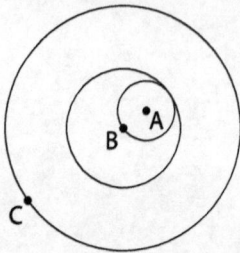

13. The centers of the three circles above lie on segment $\overline{AC}$. The center of the largest circle is point B, and the center of the smallest is point A. The middle circle and largest circle are concentric. If the middle circle bisects the radius of the largest circle, and the radius of the largest circle is 8, what is the area of the smallest circle?

(A) π
(B) 2π
(C) 3π
(D) 4π
(E) 5π

Explanation:
Radius of large circle = $8 = \overline{BC}$

Middle circle bisects $\overline{BC}$, so its radius = 4

Diameter of the small circle is the radius of the middle circle, so radius of small circle = 2

Area of circle = πr^2
Area of small circle = $\pi(2^2)$
$= 4\pi$

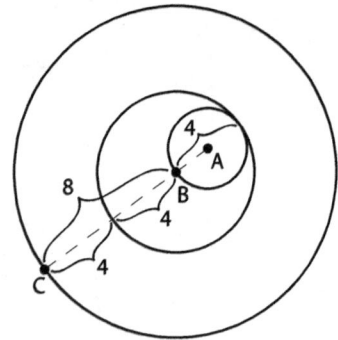

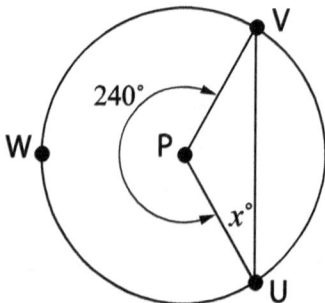

15. In circle P above, the measure of arc $\overset{\frown}{UWV}$ is 240°. What is the value of x?

(A) 30
(B) 45
(C) 60
(D) 90
(E) 120

Explanation:
When given the degree measure for an *arc*, remember that it is equal to the degree measure of the corresponding central angle.

There are 360° in a circle and $360 - 240 = 120°$, so $\angle UPV = 120°$
$\overline{PV}$ and $\overline{PU}$ are both radii of the circle and are therefore equal, which means $\triangle UPV$ is an isosceles triangle.

$180 - 120 = 60 \div 2 = 30°$

180

15. A can in the shape of a right circular cylinder has a height of 9 and a base with a diameter of 6. What is the distance from the center of the base to a point on the circumference of the top of the can?

(A) 8.48528
(B) 8.58629
(C) 9.48683
(D) 9.63542
(E) 10.81665

Explanation:
Draw a right circular cylinder:

The distance is the hypotenuse of a right triangle whose sides are 3 and 9.

$3^2 + 9^2 = c^2$
$9 + 81 = c^2$
$90 = c^2$
$\sqrt{90} = \sqrt{c^2}$
$c = \sqrt{90} = 9.48683$

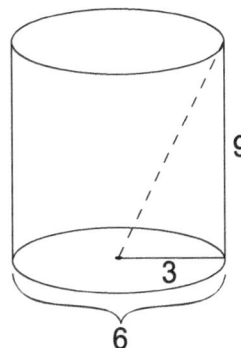

9

3

6

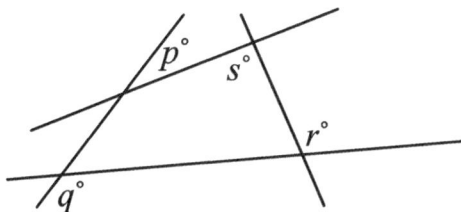

$p°$ $s°$ $r°$ $q°$

15. Four lines intersect as shown in the figure above. If $s = 100$, what is the sum of $p, q,$ and r?

(A) 180
(B) 200
(C) 280
(D) 360
(E) It cannot be determined from the given information.

Explanation:
We have a Sneaky Plug In. They already give us $s = 100$, so plug in for the three remaining interior angles of the quadrilateral. Let's say the angle adjacent to r is 130°, the angle adjacent to q is 100°, and the angle adjacent to p is 30°. You can throw in any degree measures you want, as long as the four angles together add up to 360°.
$100 + 130 + 100 + 30 = 360$.

There are 180° in a straight line:
$180 - 130 = 50 = r$
$180 - 100 = 80 = q$
$180 - 30 = 150 = p$

$p + q + r =$
$150 + 80 + 50 = 280$

Nancy's unsold Girl Scout cookies

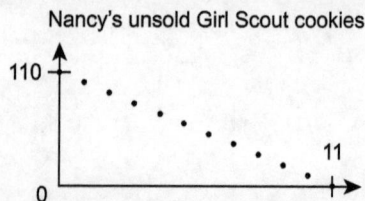

16. The number of Nancy's unsold Girl Scout cookies in an eleven day period is shown in the graph above. Which of the following equations best represents Nancy's unsold cookies?

(A) $y = 11x + 110$
(B) $y = 11x - 110$
(C) $y = 10x + 110$
(D) $y = 110 - 11x$
(E) $y = 110 - 10x$

Explanation:
The y intercept of the line is a positive 110, so eliminate (B).

The slope is negative because the graph slants down and to the right, so eliminate (A) and (C).

Rise is equal to -110 and run equal to 11

Slope: $\dfrac{-110}{11} = -10$

Slope formula: $y = mx + b$, where m = slope

So, $y = -10x + 110$, or Answer (E) $y = 110 - 10x$ works!

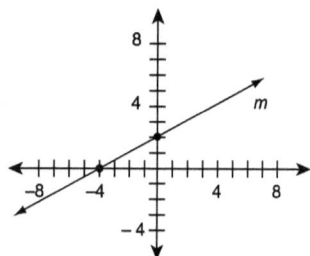

16. In the xy-plane, the equation of line m is $2y - x = 4$. Which of the following is an equation of a line that is perpendicular to line m ?

(A) $y = 2x + 6$
(B) $y = -2x + 6$
(C) $y = \frac{1}{2}x - 6$
(D) $y = \frac{1}{2}x + 4$
(E) $y = x + 4$

Explanation:
Rewrite $2y - x = 4$ into $y = mx + b$ form:
$2y - x = 4$
$2y - x + x = 4 + x$
$2y = x + 4$
$\dfrac{2y}{2} = \dfrac{x + 4}{2}$
$y = \dfrac{1}{2}x + 2$

Perpendicular lines have negative reciprocal slopes, so a line perpendicular to m would have a slope of -2.

(B) is the only option with a slope of -2!

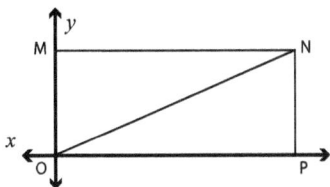

16. In the xy-plane above MNPO is a rectangle, △ONP has an area of 20, and the coordinates of P are $(10, 0)$. What is the slope of $\overline{ON}$?

(A) $\frac{3}{5}$

(B) 3

(C) $\frac{2}{5}$

(D) 5

(E) $\frac{2}{3}$

Explanation:

They tell us the coordinates of P are (10,0) which means what they are really telling us is that $\overline{OP}$, which is the base of MNOP, is 10.

Area of △ONP $= \frac{1}{2}bh$

$\frac{1}{2}(10)h = 20$

$5h = 20$

$\frac{5h}{5} = \frac{20}{5}$

$h = 4$

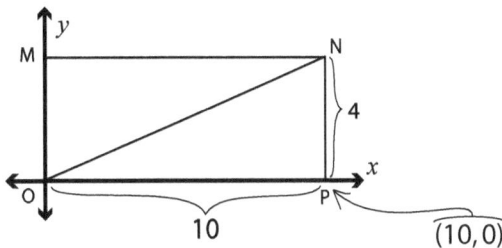

Use the origin $(0, 0)$ and coordinates of N $(10, 4)$ to find the slope using slope formula:

$\frac{y_2 - y_1}{x_2 - x_1} = slope$

$\frac{4 - 0}{10 - 0} = \frac{4}{10} = \frac{2}{5}$

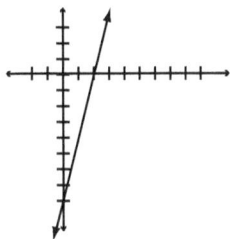

17. The equation of the line on the xy-coordinate graph is $y = 4x - 8$. Which of the following is the graph of $y = |4x - 8|$?

(A)

(B)

(C)

(D)

(E)

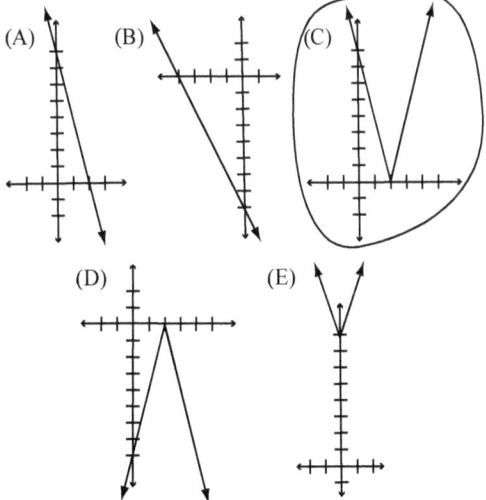

Explanation:

Plug in for x and solve for y and plot the graph.

For example:

$x = -1$

$y = |4x - 8|$

$y = |4(-1) - 8|$

$y = |-4 - 8|$

$y = |-12|$

$y = 12$

x	y
−1	12
0	8
1	4
2	0
3	4
4	8

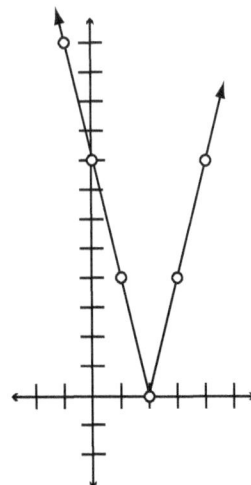

183

17. A trough in the shape of a rectangular solid has interior measurements of length 7 inches, width 3π inches, and height 4 inches. This trough is completely filled with water. All of the water is then poured into a container with an interior in the shape of a right circular cylinder with a radius of 4 inches. What must be the minimum inside height of the container?

(A) $\frac{4}{3}$

(B) $\frac{7}{3}$

(C) 2

(D) 4

(E) $\frac{21}{4}$

Explanation:
Find the volume of the rectangular solid:
$l \bullet w \bullet h = $ volume
$7 \times 3\pi \times 4 = 84\pi$

84π has to fit inside the cylindrical container. Set the formula for volume of a right circular cylinder equal to 84π, plugging in 4 for the radius.

$\pi r^2 h = $ volume
$\pi 4^2 h = 84\pi$
$16\pi h = 84\pi$
$\dfrac{16\pi h}{16\pi} = \dfrac{84\pi}{16\pi}$
$h = \dfrac{21}{4}$

$h=4$

$l=7$ $w=3\pi$

$r=4$

x

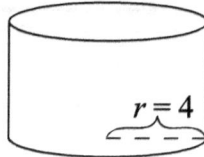

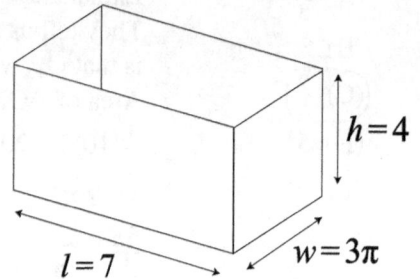

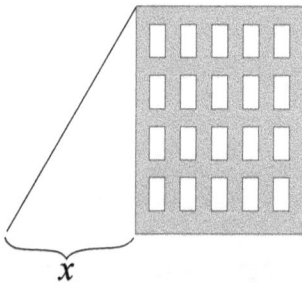

18. A ladder is extended from the roof of a building to the ground. The building is 1752.5 feet tall and the ladder has a slope of $\frac{5}{2}$. What is x, in feet?

(A) 350.5

(B) 701

(C) 876.25

(D) 1402

(E) 3505

Explanation:
Perhaps you noticed that there are numbers in the answer choices and the question asks for a single specific value, x. Use the A.C.T. Plug in for x, starting with (C).

(C) 876.25
$$\frac{\text{rise}}{\text{run}(x)} = \frac{1752.5}{876.25} = 2 \neq \frac{5}{2} \quad \times$$

(B) 701
$$\frac{\text{rise}}{\text{run}(x)} = \frac{1752.5}{701} = \frac{5}{2} = \frac{5}{2} \quad \checkmark$$

Also, it can be solved algebraically using a proportion:

$$\frac{5}{2} = \frac{1752.5}{x}$$

$$5x = 3505$$

$$\frac{5x}{5} = \frac{3505}{5}$$

$$x = 701$$

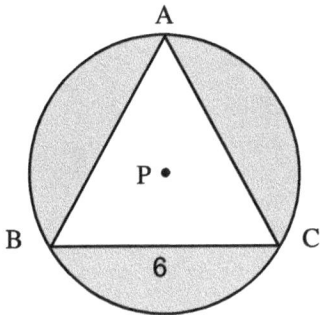

20. In the figure above, equilateral triangle ABC is inscribed in the circle with center P. What is the area of the shaded portion of the circle?

(A) $\frac{9\pi}{2} - 9\sqrt{3}$

(B) $\frac{9\pi}{2} - \frac{9\sqrt{3}}{2}$

(C) $12\pi - 9\sqrt{3}$

(D) $12\pi - \frac{9\sqrt{3}}{2}$

(E) $\frac{27\pi}{2} - \frac{9\sqrt{3}}{2}$

Explanation:
First, break ABC into two 30–60–90 triangles to find the triangle's height:

short to medium: multiply by $\sqrt{3}$
$3 \times \sqrt{3} = 3\sqrt{3}$
$h = 3\sqrt{3}$

Using the height, we can calculate area:
area of triangle $= \frac{1}{2}bh$
$b = 6, h = 3\sqrt{3}$
area $= \frac{1}{2} \times 6 \times 3\sqrt{3}$
area $= 9\sqrt{3}$

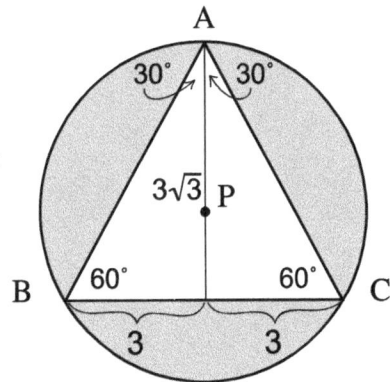

From here, we can eliminate
Answer choices (B), (D), and (E)
Now draw another 30-60-90 triangle:

medium side = 3
medium to short: divide by $\sqrt{3}$
short side $= \frac{3}{\sqrt{3}}$

short to long: multiply by 2
long side $= 2 \times \frac{3}{\sqrt{3}} =$
$r = \frac{6}{\sqrt{3}}$

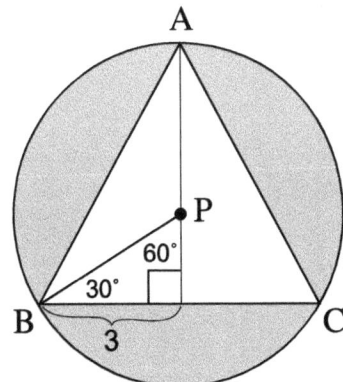

Now determine the area of the circle, using the radius:

area of circle = πr^2

area of circle = $\pi\left(\dfrac{6}{\sqrt{3}}\right)^2$

$\qquad\qquad = \pi\left(\dfrac{6}{\sqrt{3}} \cdot \dfrac{6}{\sqrt{3}}\right) = \pi\left(\dfrac{36}{3}\right) = 12\pi$

Shaded area = area of circle − area of triangle

Shaded area = $12\pi - 9\sqrt{3}$

Chapter 10
Functions Lesson

You know all of those $f(x)$, $g(x)$, and $h(x)$ problems that are all over the SAT? These are function problems, and they tend to scare students away. I see students skip these problems more than any other kind of problem on the SAT, but once you get the hang of functions and understand what is actually going on and being asked, they become some of the least complicated problems on the test. It's all about plugging in!

Functions are simply a way to find ordered pairs. Functions express a relationship between x and y points.

The first thing to know is that $f(x) = y$. So if ETS says $f(x) = 10$, they are telling you that y is 10. $g(x) = -8$ means $y = -8$. If ETS says $f(3) = 5$ they are actually giving you an ordered pair where $x = 3$ and $y = 5$.

So if we see $g(2) = 0$ what is ETS giving us? The ordered pair $(2, 0)$

If the question reads, "blah, blah, blah…what is $f(3)$?" ETS is really asking, what's the value of y when $x = 3$? That's easy. Go find it on your graph!

Let's try a couple:

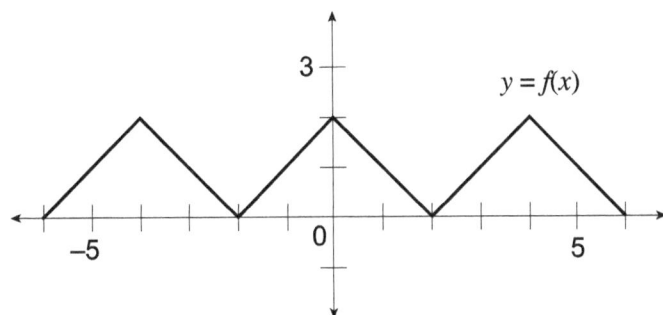

$f(3) = ?$ answer: 1

$f(-2) = ?$ answer: 0

When $f(x) = 2$, $x = ?$ In other words, what is x when $y = 2$?

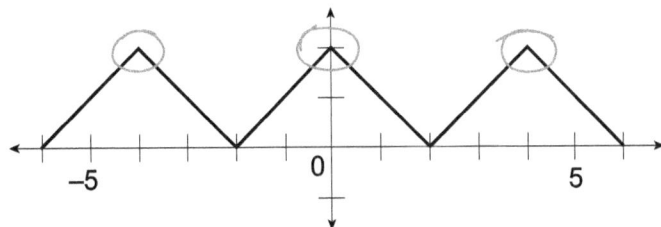

Answer: $-4, 0, 4$

Let's see how it works on an SAT problem:

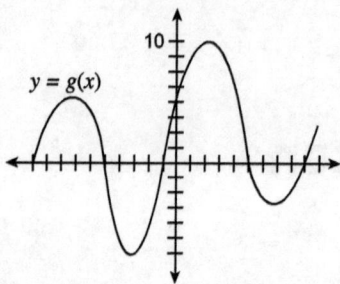

$y = g(x)$

16. The graph of $y = g(x)$ is represented above. For what value of x in the interval from $x = -10$ to $x = 10$ does the function of g attain its minimum value?

(A) −2
(B) −3
(C) −7
(D) −9
(E) −10

Explanation:

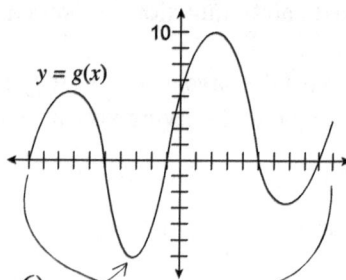

$y = g(x)$

lowest point: $(-3, -6)$

On the interval from $x = -10$ to $x = 10$, y is lowest when $x = -3$.

Answer: (B) −3

---

12. Which of the following graphs shows a function g such that $g(x) = 3$ for exactly two values of x between −6 and 6?

(A)
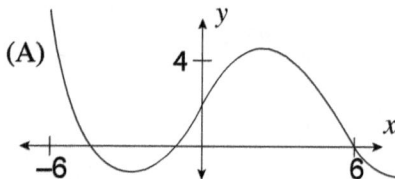

(B)
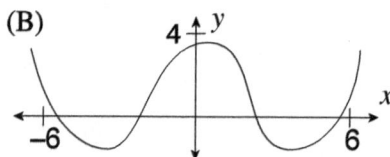

(C)
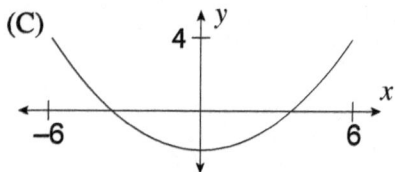

(E)
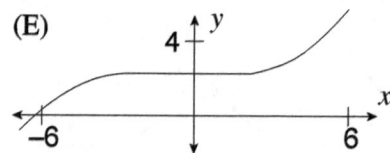

(D)
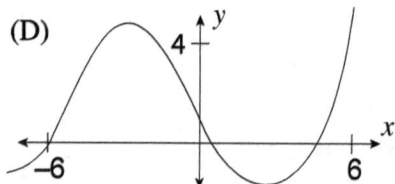

Explanation:
$g(x) = 3$ means that $y = 3$. When we plot the horizontal line $y = 3$, we count how many times this crosses the given graphed functions. Regardless of where we place the horizontal line for 3 on the y-axis (below the given 4 mark), the graph would cross that line exactly twice only in (C).

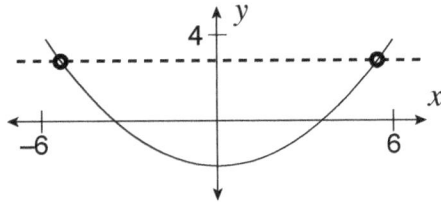

In (A), the graph would cross the line three times.

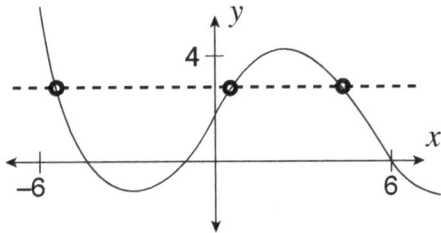

In (B), it would cross four times.

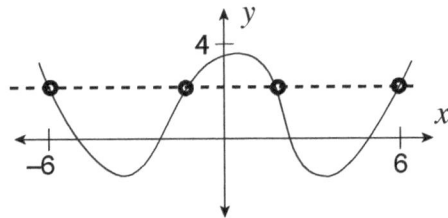

In (D), it would cross three times.

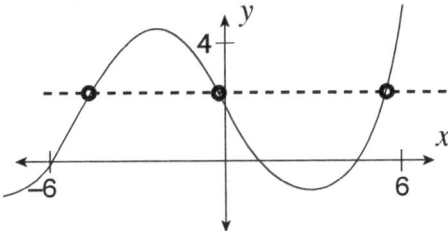

In (E), it would cross either once or infinitely many times.

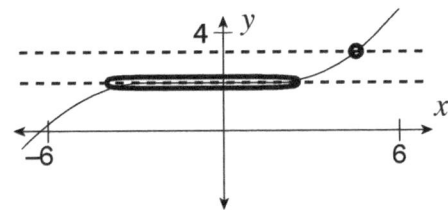

Answer: (C)

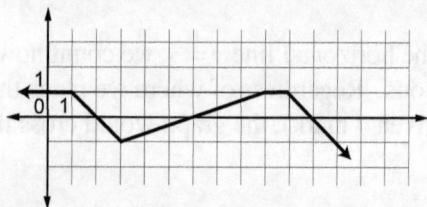

18. The figure above shows the graph of $g(x) = y$, where g is a function. If $g(z) = g(3z)$, which of the following could be the value of z?

(A) 1
(B) 2
(C) 3
(D) 4
(E) 5

Explanation:
Plug in the answer choices, starting with (C) 3:
$g(z) = g(3z)$
$g(3) = g(3 \times 3)$
$g(3) = g(9)$

On the graph, when $x = 3$, $y = -1$ and when $x = 9$, $y = 1$, so they are not equal.

Now try (B) 2:
$g(z) = g(3z)$
$g(2) = g(3 \times 2)$
$g(2) = g(6)$

On the graph, when $x = 2$, $y = 0$ and when $x = 6$, $y = 0$, so they are equal.

Answer: (B) 2

STANDARD FUNCTIONS

Many function problems on the SAT do not involve graphs at all. Usually, all ETS wants you to do is to plug in the number inside the parentheses for x and solve. For example, let's say ETS gives the function definition $f(x) = 3x^2 + 10x$ and then asks you to solve for $f(4)$. All you need to do is replace each x with a 4.

$$3x^2 + 10x =$$
$$3(4)^2 + 10(4) =$$
$$3(16) + 10(4) =$$
$$48 + 40 = 88$$

It's that easy! Let's try some SAT problems:

9. Let the function f be defined by $f(x) = 3x - 4y$, where y is a constant. If $f(7) + f(9) = 88$, what is the value of y?

(A) −20
(B) −5
(C) 0
(D) 5
(E) 20

190

Explanation:

Step 1:	Step 2:	Step 3:

$$f(7) = 3(7) - 4y$$
$$= 21 - 4y$$

$$f(9) = 3(9) - 4y$$
$$= 27 - 4y$$

$$f(7) + f(9) = 88$$
$$21 - 4y + 27 - 4y = 88$$
$$48 - 8y = 88$$
$$48 - 8y - 48 = 88 - 48$$
$$-8y = 40$$
$$\frac{-8y}{-8} = \frac{40}{-8}$$
$$y = -5$$

Answer: (B) −5

---

13. A population of parrots was observed in a controlled rain forest setting. The function $h(s) = 32(6)^{s/4}$ expresses the population of parrots s months after observation began. According to this function, what is the population of parrots 8 months after observation began?

(A) 1152
(B) 1251
(C) 1512
(D) 2511
(E) 2512

Explanation:

$$h(s) = 32(6)^{s/4}$$
$$s = 8$$
$$h(8) = 32(6)^{8/4}$$
$$= 32(6)^2$$
$$= 32(36)$$
$$= 1152$$

Answer: (A) 1152

FUNKY FUNCTIONS

Funky functions are just like the standard $f(x)$ functions. You know you are on a funky function problem when you see a weird symbol and your first reaction is, "what the heck is that? I never learned that!" All you need to do is to plug whatever they give you for the variables into the function given in the definition.

Remember: If ETS makes up a symbol, they are trying to make the problem look artificially hard. Don't be spooked; give it a shot!

For example, let's say they give us the function definition: $\boxed{x} = 2x + 10$ and ask for the value of $\boxed{3}$. $\boxed{x}$ is just another way of saying $f(x)$. Simply plug 3 into the x spot.

$$\boxed{x} = 2x + 10$$
$$2(3) + 10 =$$
$$6 + 10 = 16$$

Be careful – ETS might be sneaky and put functions in the answer choices. $\boxed{16}$ is not the same as the number 16.

If functions are in the answer choices you have to solve those functions to see which one gives you 16.

What if you see something like this? $\boxed{t} = 4$ This reads as, "the function of t equals 4."

Careful! Don't plug 4 in for x. 4 isn't in the box, t is! So plug t in for x, set the whole thing equal to 4, and solve for t.

$$2t + 10 = 4$$
$$2t + 10 - 10 = 4 - 10$$
$$2t = -6$$
$$\frac{2t}{2} = -6$$
$$t = -3$$

Let's try some:

16. For all numbers r and s, let $r \Diamond s$ be defined as
$r \Diamond s = 2r^2 + s^2$. What is the value of $5 \Diamond (1 \Diamond 5)$?

(A) 86
(B) 135
(C) 729
(D) 779
(E) 1483

Explanation:
This question gives us a Funky Function with two inputs (r and s), but the same function rules apply. Solve for the parentheses ($1 \Diamond 5$) first:
$r \Diamond s = 2r^2 + s^2$
$1 \Diamond 5 = 2(1)^2 + 5^2$
$= 2 + 25$
$= 27$

Now, substitute 27 in for $1 \Diamond 5$
$5 \Diamond 27 = 2(5)^2 + 27^2$
$= 50 + 729$
$= 779$

Answer: (D) 779

---

14. Let $k \blacklozenge$ be defined as $(k + 2)(k - 2)$, for all positive integers k. Which of the following is equal to $4 \blacklozenge + 3 \blacklozenge$?

(A) $9 \blacklozenge - 8 \blacklozenge$
(B) $8 \blacklozenge - 7 \blacklozenge$
(C) $8 \blacklozenge - 6 \blacklozenge$
(D) $7 \blacklozenge - 6 \blacklozenge$
(E) $7 \blacklozenge - 5 \blacklozenge$

Explanation:
$k \blacklozenge = (k + 2)(k - 2)$
$4 \blacklozenge + 3 \blacklozenge = (4+2)(4 - 2) + (3+2)(3 - 2)$
$= (6)(2) + (5)(1)$
$= 12 + 5$
$= 17$

$\boxed{17}$

There are functions in the answer choices, so we need to work them out until we find the one that equals 17.

Plug in (C) $8 \blacklozenge - 6 \blacklozenge$
$8 \blacklozenge - 6 \blacklozenge = (8+2)(8 - 2) - (6+2)(6 - 2)$
$= (10)(6) - (8)(4)$
$= 60 - 32$
$= 28 \neq 17$

Plug in (B) $8 \blacklozenge - 7 \blacklozenge$
$8 \blacklozenge - 7 \blacklozenge = (8+2)(8 - 2) - (7+2)(7 - 2)$
$= (10)(6) - (9)(5)$
$= 60 - 45$
$= 15 \neq 17$

Plug in (A) $9\spadesuit - 8\spadesuit$

$9\spadesuit - 8\spadesuit = (9+2)(9-2) - (8+2)(8-2)$

$= (11)(7) - (10)(6)$

$= 77 - 60$

$= 17$

Answer: (A) $9\spadesuit - 8\spadesuit$

GRAPHING PARABOLAS

No function lesson is complete without discussing transformations and their effects on graphs. Typically, we can plug in x and y values to see how a particular graph will be transformed, but when dealing with parabolas it is so much easier to just know your parabola rules.

The basic equation for a parabola is $y = x^2$.

It looks like this:

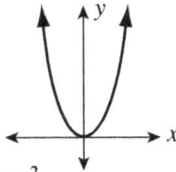

When a negative is placed in front of the x^2, the graph flips upside down.

$y = -x^2$ looks like this:

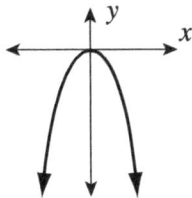

When a fraction less than one is placed in front of the x^2, the graph becomes wider.

$y = \frac{2}{5}x^2$ looks like this:

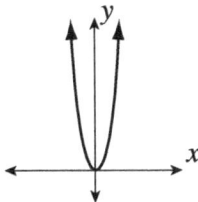

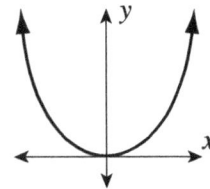

When a whole number is placed in front of the x^2, the graph becomes narrower.

$y = 5x^2$ looks like this:

When a number is added to the x^2, the graph shifts up.

$y = x^2 + 3$ looks like this:

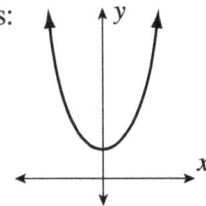

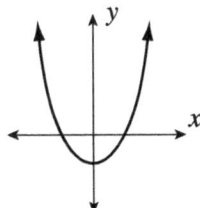

When a number is subtracted from the x^2, the graph shifts down.

$y = x^2 - 3$ looks like this:

When the x is placed in parentheses and a number is added or subtracted and that whole quantity is squared, the graph shifts to the right or the left (in the opposite direction of the symbol).

$y = (x + 2)^2$ looks like this:

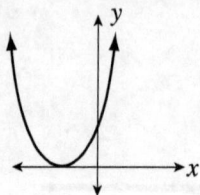

+ 2 moves it LEFT 2

$y = (x - 2)^2$ looks like this:

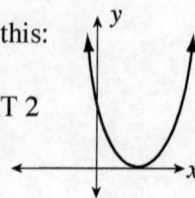

− 2 moves it RIGHT 2

Here are some more examples:

$y = x^2 - 2$

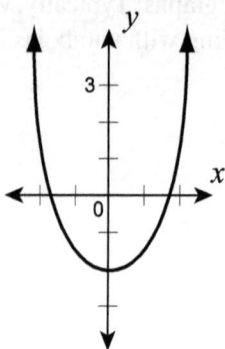

$y = \frac{1}{2}x^2 + 4$

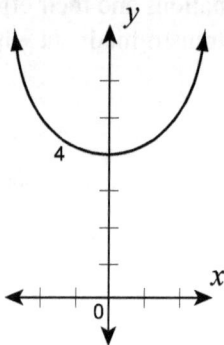

$y = -(x + 2)^2 - 3$

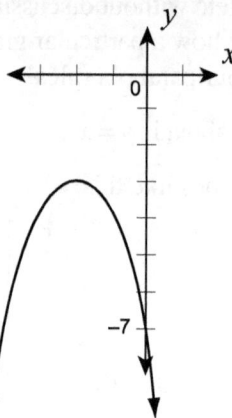

$y = 3x^2 + \frac{1}{2}$

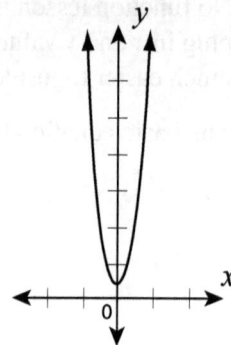

These rules can be used for the graphs of other functions, including lines.

Let's try an SAT problem:

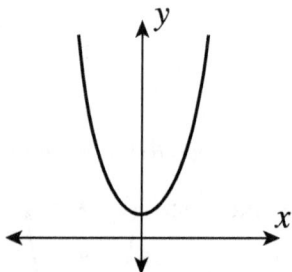

15. The equation of the parabola above is $y = bx^2 + 4$, where b is a constant. If $y = 2bx^2 + 4$ is graphed on the same axes, which of the following best describes the resulting graph when compared to the graph above?

(A) It will be moved to the left.
(B) It will be moved to the right.
(C) It will be moved 2 units downward.
(D) It will be wider.
(E) It will be narrower.

Explanation:
We can always plug in for b and then see what happens to y when we plug in different x values. However, it's a lot easier and a lot less time consuming if we just know our parabola rules. When a whole number appears before the x^2 term the parabola becomes narrower.

Answer: (E) It will be narrower.

Time for a drill of 10 function problems!

Functions Drill

6. For all positive integers k, let $k\downarrow$ be defined as the cube of the smallest prime factor of k. What is the value of $12\downarrow$?

(A) 2
(B) 3
(C) 8
(D) 27
(E) 81

x	$g(x)$
–2	1
–1	–2
0	–1
1	2
2	1

8. A portion of the function g is defined by the table above. For what value of x does $2x - g(x) = 3$?

(A) –2
(B) –1
(C) 0
(D) 1
(E) 2

10. If the function h is defined by $h(x) = 5x + 8$, then $3h(x) + 7 =$

(A) $8x + 8$
(B) $8x + 16$
(C) $15x + 4$
(D) $15x + 16$
(E) $15x + 31$

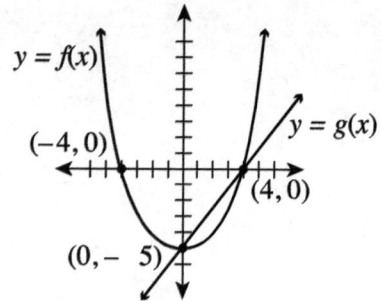

15. The figure above shows portions of the graphs of the functions f and g. What are all values of x between –6 and 6 for which $f(x) < g(x)$?

(A) $-6 < x < -4$ only
(B) $0 < x < 4$ only
(C) $-4 < x < 0$ only
(D) $4 < x < 6$ only
(E) $-6 < x < -4$ and $0 < x < 4$

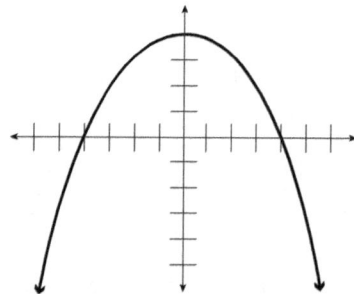

16. The graph above represents the function h, where $h(x) = c(x + 4)(x - 4)$ for some constant c. If $h(d - 2.3) = 0$ and $d > 0$, what is the value of d?

(A) 2.7
(B) 3
(C) 4
(D) 6.3
(E) 4.3

16. In the xy-plane, the graph of the function h is a line. If $h(3) = 6$, and $h(9) = 2$, what is the value of $h(6)$?

(A) 6.2
(B) 6
(C) 5
(D) 4.6
(E) 4

x	$f(x)$
−1	3
0	2
1	4
2	0
3	−3
4	−1
5	5

17. The table above shows several values of the function f. The function h is defined by $h(x) = f(2x - 3)$. What is the value of $h(3)$?

(A) −3
(B) −1
(C) 0
(D) 2
(E) 3

$$m(c) = 300(0.64)^c$$

17. The function above can be used to demonstrate the population of a certain species of freshwater fish in the Great Lakes. If $m(c)$ gives the number of the species living c decades after the year 1989, which of the following is true about the population of the fish from 1989 to 2009?

(A) It increased by 300
(B) It increased by 120
(C) It decreased by 180
(D) It decreased by 120
(E) It decreased by 300

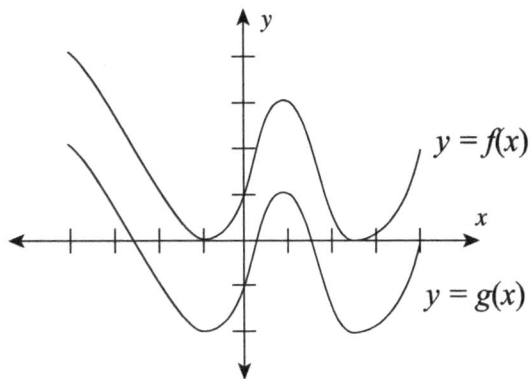

17. The figure above represents the graphs of the functions f and g in the interval from $x = -4$ to $x = 4$. Which of the following best expresses g in terms of f?

(A) $f(x) + 2$
(B) $f(x + 2)$
(C) $f(x + 2) + 2$
(D) $f(x) - 2$
(E) $f(x - 2)$

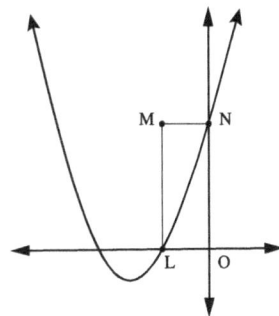

18. The graph of $y = 3x^2 + 10x + 8$ intersects the y axis at N and the x axis at L, as shown in the figure above. What is the area of rectangle LMNO?

(A) 26
(B) $\frac{33}{2}$
(C) 35
(D) $\frac{32}{3}$
(E) 33

Answers and Explanations

<table>
<tr><td colspan="3">Answer Key:</td></tr>
<tr><td>6. (C)</td><td>16. (D)</td><td>17. (D)</td></tr>
<tr><td>8. (E)</td><td>16. (E)</td><td>18. (D)</td></tr>
<tr><td>10. (E)</td><td>17. (A)</td><td></td></tr>
<tr><td>15. (B)</td><td>17. (C)</td><td></td></tr>
</table>

6. For all positive integers k, let $k\downarrow$ be defined as the cube of the smallest prime factor of k. What is the value of $12\downarrow$?

(A) 2
(B) 3
(C) 8
(D) 27
(E) 81

Explanation:
Find the prime factors of 12:

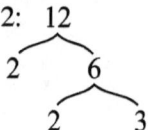

12
2   6
2   3

So, 2 is the smallest prime factor of 12. Now find the cube of 2:
$2^3 = 8$
$12\downarrow = 8$

x	$g(x)$
−2	1
−1	−2
0	−1
1	2
2	1

8. A portion of the function g is defined by the table above. For what value of x does $2x - g(x) = 3$?

(A) −2
(B) −1
(C) 0
(D) 1
(E) 2

Explanation:
Plug in for x, starting with (C) 0
$2(0) - g(0) = 3$
$0 - (-1) = 3$
$0 + 1 = 3$
$1 \ne 3$
Remember: $g(0)$ translates to "what is y when x is zero?" When x is 0, y is −1.

(D) 1
$2(1) - g(1) = 3$
According to the table, when x is 1, y is 2.
$2 - 2 = 0$
$0 \ne 3$

(E) 2
$2(2) - g(2) = 3$
According to the table, when x is 2, y is 1.
$4 - 1 = 3$
$3 = 3$ ✓

10. If the function h is defined by $h(x) = 5x + 8$,
then $3h(x) + 7 =$

(A) $8x + 8$
(B) $8x + 16$
(C) $15x + 4$
(D) $15x + 16$
(E) $15x + 31$

Explanation:
There are variables in the answer choices so it might be the safest bet to plug in.
However, if your algebra skills are sharp, it is quite easy to solve algebraically.
$h(x) = 5x + 8$
$3(5x + 8) + 7 =$
$15x + 24 + 7 = 15x + 31$

Or, by plugging in a number for x, such as 2:
$x = 2$
$h(2) = (5 \times 2) + 8$
$\quad\quad = 10 + 8 = 18$
$h(2) = 18$

$3h(2) + 7 =$
$3(18) + 7 = 61$ $\boxed{61}$

(A) $8x + 8$
$\quad 8(2) + 8 = 24$

(B) $8x + 16$
$\quad 8(2) + 16 = 32$

(C) $15x + 4$
$\quad 15(2) + 4 = 34$

(D) $15x + 16$
$\quad 15(2) + 16 = 46$

(E) $15x + 31$
$\quad 15(2) + 31 = 61$

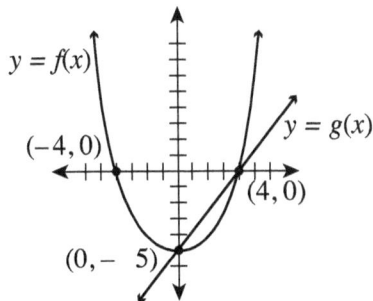

15. The figure above shows portions of the graphs
of the functions f and g. What are all values of
x between -6 and 6 for which $f(x) < g(x)$?

(A) $-6 < x < -4$ only
(B) $0 < x < 4$ only
(C) $-4 < x < 0$ only
(D) $4 < x < 6$ only
(E) $-6 < x < -4$ and $0 < x < 4$

Explanation:
Plug points within the domain given into the answer choices. Note: $g(x)$ refers to the line and $f(x)$ refers to the parabola. The parabola needs to be less than the line.

(A) $-6 < x < -4$ only

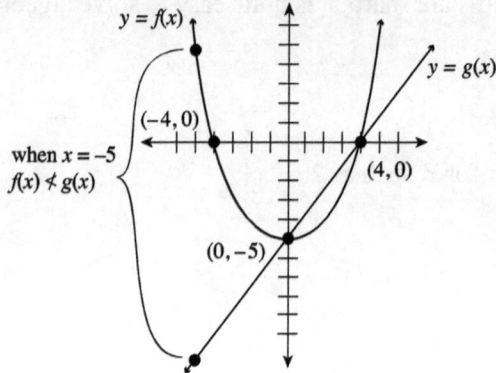

(D) $4 < x < 6$ only

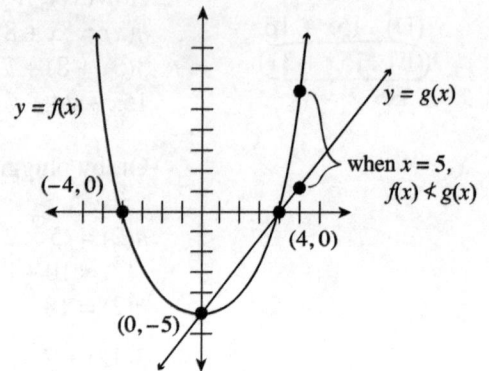

(B) $0 < x < 4$ only

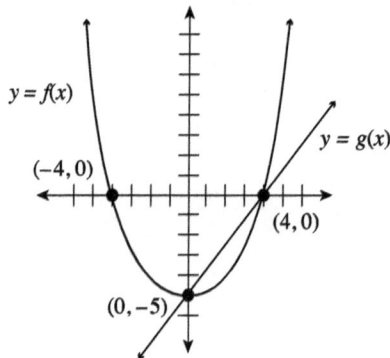

Between 0 and 4, all values of y for $g(x)$ are greater than all values for $f(x)$.

(E) $-6 < x < -4$ and $0 < x < 4$

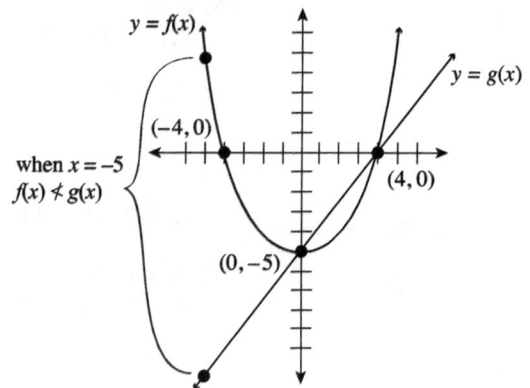

(C) $-4 < x < 0$ only

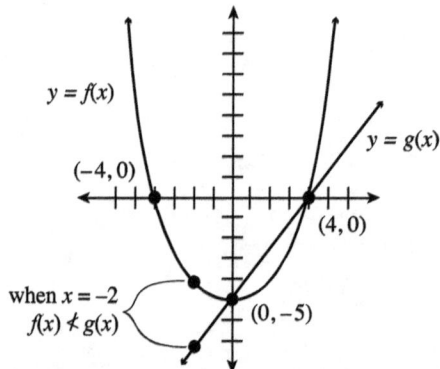

Alternatively, since $f(x)$ and $g(x)$ represent y values, we are looking for when the graph of $f(x)$ (the parabola) is below the graph of $g(x)$ (the straight line). When x is between 0 and 4, $f(x)$ is below $g(x)$!

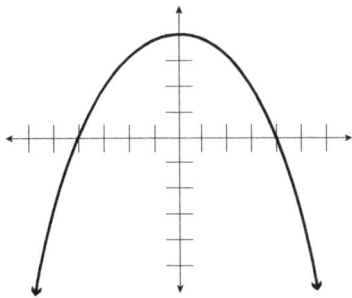

16. The graph above represents the function h, where $h(x) = c(x + 4)(x - 4)$ for some constant c. If $h(d - 2.3) = 0$ and $d > 0$, what is the value of d?

(A) 2.7
(B) 3
(C) 4
(D) 6.3
(E) 4.3

Explanation:
There is information in this problem that we don't need to use: $h(x) = c(x + 4)(x - 4)$. Sometimes ETS does this to make the problem seem harder.

Let's rewrite $h(d - 2.3) = 0$ as $h(x) = 0$.
On the graph, when $y = 0$, x could be 4 or -4, so,
$d - 2.3 = 4$
$d - 2.3 + 2.3 = 4 + 2.3$
$d = 6.3$

Or: $d - 2.3 = -4$
$d - 2.3 + 2.3 = -4 + 2.3$
$d = -1.7$
But it says $d > 0$, so (D) works!

---

16. In the xy-plane, the graph of the function h is a line. If $h(3) = 6$, and $h(9) = 2$, what is the value of $h(6)$?

(A) 6.2
(B) 6
(C) 5
(D) 4.6
(E) 4

Explanation:
Option 1:
$h(3) = 6$ is the coordinate point $(3, 6)$ and $h(9) = 2$ is $(9, 2)$.
Plot the points on a graph and visually ballpark your y when x is 6.

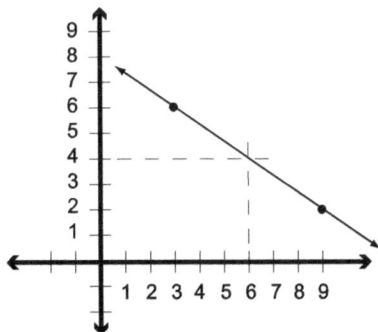

It is around 4 or so and you can definitely eliminate (A) and (B).

Option 2:
Step 1:
Find the slope (m) using the two given points:
$$\frac{y_2 - y_1}{x_2 - x_1} = m$$
$(3,6)\ (9,2)$
$$\frac{2 - 6}{9 - 3} = m$$
$$m = \frac{-4}{6} = \frac{-2}{3}$$

Step 2:
$$\frac{y_2 - y_1}{x_2 - x_1} = m$$
$(3,6)(6,y)$
$$\frac{y - 6}{6 - 3} = \frac{-2}{3}$$
$$\frac{y - 6}{3} = \frac{-2}{3}$$
$$3(y - 6) = -6$$
$$3y - 18 = -6$$
$$3y - 18 + 18 = -6 + 18$$
$$3y = 12$$
$$\frac{3y}{3} = \frac{12}{3}$$
$$y = 4$$

201

x	$f(x)$
−1	3
0	2
1	4
2	0
3	−3
4	−1
5	5

17. The table above shows several values of the function f. The function h is defined by $h(x) = f(2x - 3)$. What is the value of $h(3)$?

(A) −3
(B) −1
(C) 0
(D) 2
(E) 3

Explanation:
They are giving us that $x = 3$ so start by plugging in 3 for x.

$$h(x) = f(2x - 3)$$

$$h(3) = f(2(3) - 3)$$

$$h(3) = f(6 - 3)$$

$$h(3) = f(3)$$

Go to the table and figure out what y is when x is 3.

$$h(3) = -3$$

x	$f(x)$
−1	3
0	2
1	4
2	0
3	−3
4	−1
5	5

$$m(c) = 300(0.64)^c$$

17. The function above can be used to demonstrate the population of a certain species of freshwater fish in the Great Lakes. If $m(c)$ gives the number of the species living c decades after the year 1989, which of the following is most true about the population of the fish from 1989 to 2009?

(A) It increased by 300
(B) It increased by 120
(C) It decreased by 180
(D) It decreased by 120
(E) It decreased by 300

Explanation:
To find the original population, set $c = 0$.
$$m(c) = 300(.64)^0$$
$$= 300(1)$$
$$= 300$$

Set $c = 2$ because there are two decades from 1989 – 2009
$$m(c) = 300(.64)^2$$
$$= 300(.4)$$
$$= 120$$
Now find the difference:
$$300 - 120 = 180$$

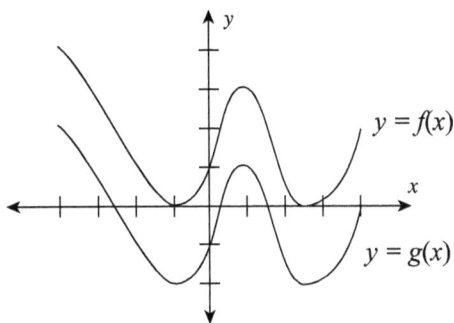

17. The figure above represents the graphs of the functions f and g in the interval from $x = -4$ to $x = 4$. Which of the following best expresses g in terms of f?

(A) $f(x) + 2$
(B) $f(x + 2)$
(C) $f(x + 2) + 2$
(D) $f(x) - 2$
(E) $f(x - 2)$

Explanation:
$g(x)$ has retained the general shape of $f(x)$ and has shifted down approximately 2. So, $g(x) = f(x) - 2$.

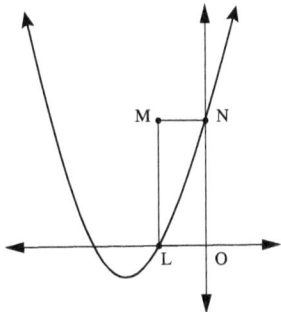

18. The graph of $y = 3x^2 + 10x + 8$ intersects the y axis at N and the x axis at L, as shown in the figure above. What is the area of rectangle LMNO?

(A) 26
(B) $\dfrac{33}{2}$
(C) 35
(D) $\dfrac{32}{3}$
(E) 33

Explanation:
Set $x = 0$ and solve for y to find the point on the graph where the parabola crosses the y axis:
$y = 3x^2 + 10x + 8$
$y = 3(0) + 10(0) + 8$
$y = 8$

Now, set y equal to 0 to find the point on the graph where the parabola crosses the x axis:
$y = 3x^2 + 10x + 8$
$3x^2 + 10x + 8 = 0$
$(3x + 4)(x + 2) = 0$
$3x + 4 = 0$ or $x + 2 = 0$

$3x + 4 = 0$

$3x + 4 - 4 = 0 - 4$

$3x = -4$

$\dfrac{3x}{3} = \dfrac{-4}{3}$

$x = -\dfrac{4}{3}$

or $x + 2 = 0$

$x + 2 - 2 = 0 - 2$

$x = -2$

$x = -\dfrac{4}{3}$ or $x = -2$

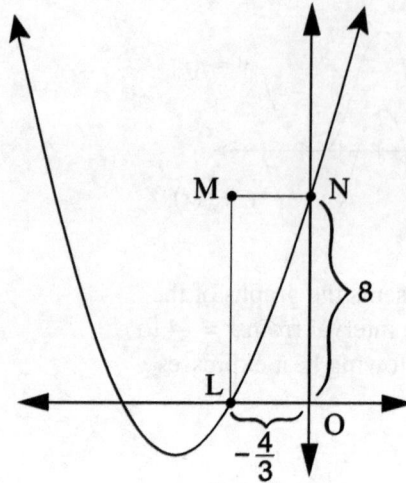

Since L is the intersection closest to 0, we know that the x coordinate of point L is $-\dfrac{4}{3}$

Area of LMNO $= \dfrac{4}{3} \times 8$

$= \dfrac{32}{3}$

Conclusion

You've learned a lot. I've thrown a lot of information at you, and it may take a while to digest. Practice the techniques, learn the material, and start working through actual SAT tests. Get a copy of the College Board *Official SAT Study Guide* and work through the practice tests. They don't give explanations to the problems, but I have given you the knowledge and the skills to work through every math problem in that guide. And remember: the problems you find too challenging, you get to leave blank!

Take the SAT more than once. The first time around take the SAT to get a feel for the test and to get your baseline score. Practice hard and then try to beat your previous scores the second time around. And if you want to take it a third time, that's totally cool; just keep pushing to do better.

Be sure to put new batteries in your calculator the night before test day. And bring along some extra just in case! Take a couple of number 2 pencils, water and a snack to refuel during the break, and a head full of confidence. If you find yourself confused and/or stressed, remember that you can always skip (and come back to) harder problems. In general, take a deep breath, believe in yourself, and remember the rules, definitions, and strategies I have taught you in this book. Good luck!

www.ingramcontent.com/pod-product-compliance
Lightning Source LLC
La Vergne TN
LVHW061259060426
835509LV00013B/1486